T0185367

SPACE TELESCOPE SCIENCE INSTITUTE

SYMPOSIUM SERIES: 15

Series Editor S. Michael Fall, Space Telescope Science Institute

THE DARK UNIVERSE:
MATTER, ENERGY AND GRAVITY

This book reviews the recent findings on the composition of the universe, its dynamics, and the implications of both for the evolution of large-scale structure and for fundamental theories of the universe. With each chapter written by a leading expert in the field, topics include Massive Compact Halo Objects, the oldest white dwarfs, hot gas in clusters of galaxies, primordial nucleosynthesis, Modified Newtonian Dynamics, the cosmic mass density, the growth of large scale structure, and a discussion of dark energy. The book is an invaluable resource for professional astronomers and graduate students in this cutting-edge area of research.

Other titles in the Space Telescope Science Institute Series.

The Dark Universe: Matter, Energy and Gravity

Proceedings of the
Space Telescope Science Institute Symposium,
held in Baltimore, Maryland
April 2–5, 2001

Edited by
MARIO LIVIO

Space Telescope Science Institute, Baltimore, MD 21218, USA

Published for the Space Telescope Science Institute

CAMBRIDGE
UNIVERSITY PRESS

CAMBRIDGE UNIVERSITY PRESS
Cambridge, New York, Melbourne, Madrid, Cape Town, Singapore,
São Paulo, Delhi, Dubai, Tokyo

Cambridge University Press
The Edinburgh Building, Cambridge CB2 8RU, UK

Published in the United States of America by Cambridge University Press, New York

www.cambridge.org
Information on this title: www.cambridge.org/9780521134033

First published 2003
-This digitally printed version 2010

A catalogue record for this publication is available from the British Library

ISBN 978-0-521-82227-5 Hardback
ISBN 978-0-521-13403-3 Paperback

Contents

Participants

AbdelSalam, Hanadi	Kapteyn Institute for Astronomy
Alam, S. M. Khairul	Ohio State University
Albrow, Michael	Space Telescope Science Institute
Alcaniz, Jailson	Universidade Federal do Rio Grande do Norte
Alcock, Charles	Lawrence Livermore National Laboratory
Allen, Ron	Space Telescope Science Institute
Anchev, Joseph	
Andrews, Thomas	
Avera, Randy	NASA/FAA
Bahcall, John	Institute for Advanced Study
Bahcall, Neta	Princeton University
Barish, Barry	California Institute of Technology
Beckwith, Steven	Space Telescope Science Institute
Benitez, Narciso	The Johns Hopkins University
Bennett, David	University of Notre Dame
Bergvall, Nils	Astronomical Observatory of Uppsala
Bernabei, Rita	Universita di Roma II "Tor Vergata"
Blakeslee, John	The Johns Hopkins University
Bludman, Sidney	DESY-T
Burg, Richard	NASA/Goddard Space Flight Center
Burke, Christopher	Ohio State University
Burkert, Andreas	Max Plancke Institut für Astronomie
Caldwell, Robert	Dartmouth College
Canzian, Blaise	U.S. Naval Observatory
Carpenter, Kenneth	NASA/Goddard Space Flight Center
Casertano, Stefano	Space Telescope Science Institute
Chaname, Julio	Ohio State University
Cheslow, Melvyn	
Chou, C. K.	National Central University
Christian, Carol	Space Telescope Science Institute
Conselice, Christopher	Space Telescope Science Institute
Daly, Ruth	Pennsylvania State University
D'Amario, James	Harford Community College
de Jong, Jelte	Kapteyn Institute
Delahaye, Franck	Ohio State University
Dell'Antonio, Ian	Brown University
Dine, Michael	University of California at Santa Cruz
Donahue, Megan	Space Telescope Science Institute
Drake, Andrew	Lawrence Livermore National Laboratory
Duerbeck, Hilmar W.	Brussels Free University (VUB)
Dvali, Gia	New York University
Fang, Taotao	Massachusetts Institute of Technology
Fazio, Giovanni	Harvard-Smithsonian Center for Astrophysics
Felten, James	NASA/Goddard Space Flight Center
Ferguson, Harry	Space Telescope Science Institute
Fields, Dale	Ohio State University
Freudling, Wolfram	ST-European Southern Observatory
Gaitskell, Richard	University of California at Berkeley

Gerhard, Ortwin	University of Basel
Gerke, Brian	Cambridge University
Giavalisco, Mauro	Space Telescope Science Institute
Glicenstein, Jean-Francos	CEA-Saclay
Godon, Patrick	Space Telescope Science Institute
Graber, James	Library of Congress
Greyber, Howard	Greyber Associates
Guimaraes, Antonio C.	Brown University
Hämmerle, Hannelore	Universität Bonn
Hansen, Bradley	Princeton University
Hartnet, Kevin	NASA/Goddard Space Flight Center
Hauser, Michael	Space Telescope Science Institute
Henriksen, Mark	University of Maryland, Baltimore County
Hoekstra, Henk	CITA
Jain, Bhuvnesh	University of Pennsylvania
Jeletic, James	NASA/Goddard Space Flight Center
Kamionkowski, Marc	California Institute of Technology
Kassin, Susan	Ohio State University
Kazanas, Demosthenes	LHEA/NASA/Goddard Space Flight Center
Kimble, Randy	NASA/Goddard Space Flight Center
King, Lindsay	University of Bonn
Kirkham, Barry	TRW/NASA/Goddard Space Flight Center
Kochanek, Chris	Harvard-Smithsonian Center for Astrophysics
Leckrone, David	NASA/Goddard Space Flight Center
Lima, Jose Ademir	Universidade Federal do Rio Grande do Norte
Lin, Yi-Hui	National Central University
Livio, Mario	Space Telescope Science Institute
Macchetto, Duccio	Space Telescope Science Institute
Maoz, Dan	Columbia University
Margon, Bruce	Space Telescope Science Institute
Marochnik, Leonid	Space Telescope Science Institute
Marshall, Jennifer	Ohio State University
Mashchenko, Sergiy	University of Montreal
Mathews, Grant	University of Notre Dame
McKernan, Barry	University College of Dublin
Medvedev, Mikhail	CITA, University of Toronto
Meylan, Georges	Space Telescope Science Institute
Miralles, Joan-Marc	Space Telescope European Coordinating Facility
Natarajan, Priyamvada	Yale University
Nelson, Cailin	IGPP/Lawrence Livermore National Laboratory
Netterfield, C. Barth	University of Toronto
Niedner, Mal	NASA/Goddard Space Flight Center
Nomoto, Ken'icki	University of Tokyo
Onken, Christopher	Ohio State University
Osmer, Patrick	Ohio State University
Paulin-Henriksson, Stephane	PCC, College of France
Peacock, John	European Southern Observatory
Perlmutter, Saul	Lawrence Berkeley Laboratory
Pringle, James	Institute of Astronomy, Cambridge
Rauscher, Bernard J.	Space Telescope Science Institute

Reid, Iain Neill	Space Telescope Science Institute
Rhodes, Jason	NASA/Goddard Space Flight Center
Richer, Harvey	University of British Columbia
Riffeser, Arno	Sternwarte München
Riess, Adam	Space Telescope Science Institute
Rosati, Piero	European Southern Observatory
Rubin, Vera	Carnegie Institute of Washington
Runyan, Marcus	California Institute of Technology
Ryden, Barbara	Ohio State University
Sahu, Kailash	Space Telescope Science Institute
Sancisi, Renzo	Kapteyn Sterrekundig Institute
Sanders, Bob	Kapteyn Sterrekundig Institute
Schild, Rudolph	Smithsonian Astrophysical Observatory
Schneider, Peter	Universität Bonn
Schreier, Ethan	Space Telescope Science Institute
Seitter, Waltraut C.	Münster University
Shanks, Tom	University of Durham
Silverberg, Robert	NASA/Goddard Space Flight Center
Sparmo, Joe	NASA/Goddard Space Flight Center
Stecher, Theodore	NASA/Goddard Space Flight Center
Steigman, Gary	Ohio State University
Steinhardt, Paul	Princeton University
Stiavelli, Massimo	Space Telescope Science Institute
Struble, Mitchell	University of Pennsylvania, Lockheed Martin
Swaters, Robert	Carnegie Institute of Washington
Tavarez, Maritza	University of Michigan
Tinker, Jeremy	Ohio State University
Tripp, Todd	Princeton University
Turner, Michael	University of Chicago
Tyson, Anthony	Bell Laboratories, Lucent Technology
Urry, C. Megan	Space Telescope Science Institute
Vilenkin, Alex	Tufts University
Williams, Bob	Space Telescope Science Institute
Wilson, Gillian	Brown University
Woodgate, Bruce	NASA/Goddard Space Flight Center
Xie, Gaofeng	Purple Mountain Observatory
Yamamoto, Kazuhiro	Hiroshima University
Yaqoob, Tahir	The Johns Hopkins University
Zheng, Zheng	Ohio State University

Preface. Through a glass, darkly

The planet Uranus was discovered in 1781 by the British astronomer William Herschel. Not long after its discovery, astronomers charting the orbit of Uranus found small discrepancies between the predicted and observed positions of the planet. In September 1845, British astronomer John Adams proved mathematically that the deviations in Uranus' orbit could not result merely from the gravitational pull of the other known planets and he predicted the existence of another, previously undetected planet in the solar system. The eventual discovery of the planet Neptune in September 1846 by the German astronomer Johann Galle thus marked the first detection of astronomical "dark matter" whose presence was first deduced by its gravitational effects. However, in the history of physics, we also find a case in which the assumption about the existence of an unseen medium was later proven to be totally wrong. Until 1887, physicists assumed that *aether*—a substance that pervades all space—was a necessary medium for the propagation of light. A famous experiment by American researchers Albert Michelson and Howard Morley not only showed unambiguously that this medium does not exist, but the experimental results also set Einstein on the road to a new theory of space and time—special relativity.

Astrophysicists today are faced with a similar "Neptune vs. aether" dilemma. On the face of it, there are many indications that about 90% of the *matter* in our universe is in the form of "dark matter"—matter whose constituents do not emit electromagnetic radiation and that interact very weakly with ordinary matter. The luminous galaxies we see are just like the tiny minilights on a huge, dark, Christmas tree. The existence and amount of the "dark matter" is deduced, for example, from the speeds of galaxies in clusters of galaxies. In equilibrium, the gravitational force of all the matter in the cluster exactly balances the proneness of the galaxies to scatter in all directions. Careful determinations of the speeds thus "weigh" the cluster. Other observations, like *gravitational lensing* the bending of light from distant sources by the cluster's gravity—also confirm that about 90% of the mass in clusters is dark.

The most likely candidates for the constituents of the dark matter are some exotic elementary particles that are relics of the very early, high-energy universe. Elementary particle theories that link fermions (that have a fractional quantum mechanical spin) and bosons (with integer spin) are known as supersymmetry (or, affectionately, SUSY) theories. Supersymmetry requires the existence of (yet undiscovered) fractional spin, neutral, massive partners to integer spin particles like the photon. The lightest members of this menagerie of SUSY particles are known as *neutralinos* and they are the leading candidates for dark matter.

However, there is another possibility, in principle, to explain the extra gravity usually attributed to dark matter. The idea behind this alternative is similar in spirit to the lesson learned from the aether. Instead of requiring the existence of an unseen medium, maybe the *theory* of gravity itself needs to be changed. One proposed modification suggests that our three-dimensional (plus time) universe with all of its elementary particles is stuck to a (three-dimensional) membrane that exists in a higher-dimensional space. Particles like protons and electrons cannot move in the extra dimensions and neither can the electromagnetic fields (a bit like electrons being confined to move along a copper wire). Gravity and its carrier—the graviton—can, on the other hand, extend and travel into the higher-dimensional space. In this model, the gravitational effects we attribute to dark matter could simply represent the gravity of matter that resides in a membrane/universe parallel to ours. Photons cannot travel throughout the extra dimension separating the parallel universes and consequently the matter in the parallel universe is necessarily "dark" to our detectors.

Luckily, experiments planned for the coming decade will be able to distinguish between the "Neptune" and "Aether" options. The Large Hadron Collider (LHC), the world's most powerful particle accelerator which is being built in Geneva, is less than a decade away from achieving the energy range (proton beams with 7-on-7 TeV) needed to discover neutralinos. The LHC could also discover particles predicted to exist by the new theories of gravity. Furthermore, the new theory predicts deviations from Newton's inverse square law at submillimeter distances. No fewer than four experiments are expected to test gravity at these small distances during the coming few months to years.

As if the existence of dark matter was not puzzling enough, since 1998 there exists strong evidence that most of the universe's total energy density is in the form of an even more mysterious "dark energy." Observing a few dozen stellar explosions known as Type Ia supernovae at redshifts of order $z \sim 0.5$–1, two teams discovered that the expansion of the universe is accelerating!

Type Ia supernovae are extremely bright (occasionally outshining an entire galaxy) events representing the complete thermonuclear disruptions of white dwarf stars. Since Type Ia supernovae are nearly perfect "standard candles" (their small deviations from a constant luminosity are well calibrated), they can be used as superb distance indicators to distances spanning half the universe's age. The expectation prior to 1998 was that distant supernovae would reveal that the universe had been expanding in the past *faster* than predicted by a simple Hubble expansion, because of the deceleration caused by gravity. Instead, the two teams found (independently) that the distant supernovae were receding *slower* than the Hubble law, implying an accelerating cosmic expansion propelled by some "dark energy." The pressure associated with this dark energy is negative, resulting in gravity becoming a *repulsive* force. The observations, together with measurements of the anisotropy of the cosmic microwave background radiation, suggest that the energy density in the dark energy is about 73% that required for a geometrically flat universe.

The precise nature of the dark energy is probably the greatest mystery of today's physics. It is generally assumed that this dark energy represents the energy associated with the physical *vacuum*. However, the value of the observed energy density is some 55 orders of magnitude smaller than that expected from supersymmetry considerations. Currently, it is not even clear if the dark energy density is constant in time, as would be expected for Einstein's "Cosmological Constant" (introduced to produce a static universe), or evolving as some uniform scalar field (dubbed "quintessence"). It is also possible, in principle, that the accelerating universe and the deduced dark energy are also manifestations of the need for a new theory of gravity.

The Space Telescope Science Institute Symposium on "The Dark Universe: Matter, Energy, and Gravity" took place during 2–5 April 2001.

These proceedings represent a part of the invited talks that were presented at the symposium, in order of presentation. We thank the contributing authors for preparing their papers.

We thank Sharon Toolan of ST ScI for her help in preparing this volume for publication.

Mario Livio
Space Telescope Science Institute
Baltimore, Maryland

A brief history of dark matter

By VERA C. RUBIN

Department of Terrestrial Magnetism, Carnegie Institution of Washington

1. Introduction

The title not withstanding, this is not a history of dark matter. Until we know what the dark matter is, we cannot know its history. Instead, this is a brief history of how astronomers converged to the view that most of the matter in the universe is dark. This paper deals principally with the early studies which helped to answer the questions "Are rotation curves flat? If so, why?" It also includes some early history in deciphering the signature of clusters of galaxies as gravitational lenses, which seems to have been little investigated. This account covers the years up to 1980; achievements since 1980 are science, not history. Several excellent, informative brief histories exist, and interested readers should see Trimble (1987, 1995) and van den Bergh (1999). We can all thank Sidney van den Bergh for correctly translating Zwicky's "dunkle (kalte) materie" as "dark (cold) matter" and finally putting to rest the myth that Zwicky called it "missing matter."

The notion that there are stars that are dark was a common one in the 18th and 19th Century. Walt Whitman's (1855) lines in *Leaves of Grass*, "The bright suns I see and the dark suns I cannot see are in their place" and Bessel's "Foundation of an Astronomy of the Invisible" (Clerke 1885 and reference therein) are early manifestations of this belief. Based on his decade-long observations of Sirius and Procyon, Bessel announced in 1844 that each was a binary star system, whose irregular motion on the sky was due to the presence of its invisible companion. When Maria Mitchell taught her students at the new (1865) Vassar Female College, one of the classes featured Dark Stars.

Even earlier, the Reverend Mr. Mitchell (1784) had imagined a star, 500 times the solar radius but of equal density, and realized that "all light emitted from such a body would be made to return toward it, by its own proper gravity." Moreover, "if there really should exist in nature [such] bodies... since their light could not arrive at us, we could have no information from light. Yet if any other luminous bodies should happen to revolve about them, we might still perhaps from the motions of these revolving bodies infer the existence of the central ones." An impressive intuition, almost 200 years before we knew of black holes.

And of course, Vincent van Gogh's "Starry Night" (1889) is surely every optical astronomer's nightmare of what the night sky would look like if the dark matter were not dark.

2. The early 20th century

Kapteyn (1922), in his efforts to study the arrangement and motion of the sidereal system, estimated "the amount of dark matter" in a universe with the sun at the center of similar ellipsoids of revolution. Using star counts and physics, he concluded that "it appears at once that this (dark) mass cannot be excessive." Kapteyn references Jeans (1919, p. 239) for a dark star determination. However, in my 1919 edition of Jeans' book, there is no mention of dark stars on the cited page 239. Instead, the only estimate I have located in a quick perusal is on page 222, where Jeans writes "these estimates evaluate

1

the density of matter in the bright stars only; the dark stars, of which it is impossible even to guess at the number, will increase the density to a quite unknown extent, so that the estimates only provide lower limits to the true density." But in a few years Jeans had changed his mind. With later work, Jeans (1922) counts "about three dark stars in the universe for every bright star." Trimble (1995) has pointed out that this range of dark matter density matches closely the range of dark matter density currently discussed. A decade later, Oort's (1932) study of the mass density in the Galactic plane also left its mark with the name "Oort limit."

Zwicky's (1933; also 1937c) analysis of the velocity dispersion for galaxies in the Coma cluster marks the beginning of the contemporary study of dark matter in the universe, albeit a slow beginning. His study, plus that of Smith (1936) for the Virgo cluster, noted that the large relative motions for individual galaxies would disrupt the clusters, unless each galaxy has a mass about 100 times the accepted mass. Zwicky also cited good evidence that clusters are not dissolving.

The discrepancy between the high galaxy masses calculated from the viral mass of the clusters, and the low masses calculated from the very inner rotation curves for five galaxies, troubled Hubble (1936). "The discrepancy seems to be real and is important," he wrote. It is not surprising that these early absorption line rotation curves extended only over the brightest nuclear regions, and were poor indicators of galaxy mass. Several decades would pass before the cluster dark matter would be associated with the flat rotation curves derived for individual galaxies.

Observations of M31's rotation by Babcock (1939) and Mayall (1951) extended major axis rotation velocities to almost 120′ from the nucleus, but exposure times were tens of hours, and spectrographs had stability problems. Interestingly, both Babcock's velocities for M31 and Humason's unpublished velocities for NGC 3115 showed the last measured point to have a rotation velocity of over 400 km s^{-1} (almost two times actual), but consequently raised questions of mass distribution.

At a symposium celebrating the dedication of the McDonald Observatory in 1939, Oort (1940) noted that "... the distribution of mass [in NGC 3115] appears to bear almost no relation to that of the light." His conclusion, "The strongly condensed luminous system appears embedded in a large more or less homogeneous mass of great density," was a clear statement of the puzzle that would grip astronomers again in the 1970s. However, it seems to have impressed few in 1940 and in the decades following.

3. Instrumentation starts catching up: Mid-century

Early radio observations of neutral hydrogen in external galaxies showed a slowly falling rotation curve for M31 (van de Hulst el al. 1957) and a flat rotation curve for M33 (Volders 1959). Because the flatness could be attributed to the side lobes of the beam, it was consequently ignored. Louise Volders must also have realized that a flat rotation curve conflicted with the value of the Oort constants for our Galaxy, which implied a falling rotation curve at the position of the sun. Jan Oort was one of her thesis professors.

With eight rotation curves available by 1959, de Vaucouleurs (1959) concluded "In all cases the rotation curve consists of a straight inner region... beyond which the rotational velocity decreases with increasing distance to the center and tends asymptotically toward Kepler's third law." From a reanalysis of the same scattered velocities for M31, Schwarzschild (1954) reached the opposite conclusion. He stated that the approximately flat rotation curve was "not discordant with the assumption of equal mass and light distribution." With the 20/20 vision of hindsight, plots of the data reveal only a scatter of points, from which no certain conclusions can be drawn.

A paper that was more in the spirit of what was to come than what had taken place was Kahn and Woltjer's (1959) investigation of the dynamical stability of the Local Group. They concluded that the Local Group must contain an appreciable amount of invisible matter. In a sense, this was a contemporary formulation of Zwicky's virial cluster problem.

Although Zwicky's cluster results were not forgotten, they only came to the forefront of astronomy in discussions of the stability of clusters. At two symposia in Santa Barbara preceding the Berkeley 1961 General Assembly, Ambartsumian (1961) had significant support for his view that clusters were explosively disintegrating (but see van den Bergh 1961; 1999). Much of the discussion centered on galaxy radial velocities which were beginning to be obtained in fairly large numbers. My notes from the symposia mention dark matter once, in connection with the disks of early type spirals.

In an effort to learn how our Galaxy "ended," my graduate students at Georgetown and I made a study of the three dimensional velocities of almost 1000 O and B stars beyond the solar circle (Rubin et al. 1962). Our 1962 conclusion, "the stellar (rotation) curve is flat, and does not decrease as is expected for Keplerian orbits," apparently influenced very few, and was not emphasized by the senior author when she returned to the problem of galaxy rotation a decade later. In my earliest Kitt Peak observing, I attempted to obtain a rotation curve for the Galaxy beyond the solar circle by observing O and B stars near the anticenter direction (Rubin 1965). It was clear that while many studies of the center of the Galaxy were underway, there was little attention being paid to the outer limits of galaxies.

The many galaxies studied by Margaret and Geoffrey Burbidge (e.g. Burbidge, Burbidge, & Prendergast 1962) generally showed an inner velocity rise; velocities were then expected to fall. For at least one galaxy, NGC 7331 (Rubin, Burbidge, & Burbidge 1965), we showed three possible density laws which extended the velocity curve beyond the turnover; one predicted rotation curve is rising slightly, one is slightly falling. This was our way of saying that we did not know what happens beyond the final measured velocities.

4. The decade of seeing is believing: The seventies

Science often advances when ideas, formerly very disparate, are united. In retrospect, it took a long time for astronomers to relate Zwicky's dark matter to the flat rotation curves for some galaxies that were beginning to attract attention. If I were to choose a date when astronomers decided that dark matter must "really" exist, I would pick 1978. In 1977, many astronomers hoped that dark matter might be avoided; by 1979 the Annual Review article by Gallagher and Faber convinced most of the remaining skeptics. Of the two questions to be answered, "Are rotation curves flat? If so, why?" we would arrive at an answer to the first.

Freeman (1970) discussed 21-cm velocity maps of a number of nearby galaxies. For NGC 300 and M33 he concluded "if [these data] are correct, there must be in these galaxies additional matter that is not detected, either optically or at 21 cm. Its mass must be as large as the mass of the detected galaxy, and its distribution must be quite different from the exponential distribution which holds for the optical galaxy." His remarks are reminiscent of Oort's (1940) concerning NGC 3115, but technology had now made it possible to measure velocities beyond the optical galaxy.

In the same year, Kent Ford and I completed our study of the velocities of emission regions in M31 (Rubin & Ford 1970), and produced a rotation curve which extended to 120′, the extent of the optical disk. The curve was flat over the last 30% of the galaxy.

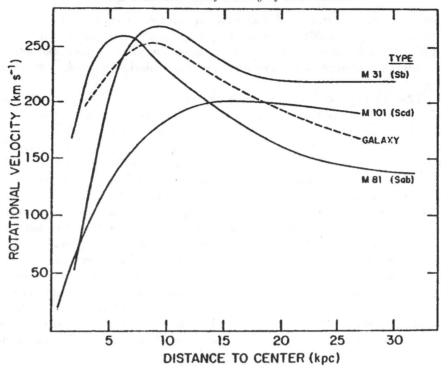

FIGURE 1. Rotation curves for the galaxies M31, M101, M81 and the rotation curve of our Galaxy, from Roberts and Rots (1973).

Perhaps that is why we chose not to employ the typical mass models that assumed a Keplerian fall beyond the last observed region, but instead wrote that "extrapolation (of the mass) beyond that distance is clearly a matter of taste." In an era when rotation curves were routinely extended in a Keplerian fashion beyond the final observed point, we chose not to do so.

More flat 21-cm rotation curves followed. Rogstad and Shostak (1972) and later Krumm and Salpeter (1976) obtained flat 21-cm rotation curves for more galaxies, but rumors of sidelobe problems continued to plague such studies. Mort Roberts, whose very extended rotation curve of M31 (1995) would lead to the acceptance of flat rotation curves, delayed that acceptance by publishing (Roberts & Rots 1973) a plot (Fig. 1) of superposed rotation curves for M31, M101, our Galaxy, and M81. For our Galaxy and M81 the outer velocity decreases were so substantial that the eye of the beholder remembered mostly the falling parts.

However, the theorists had their eyes wide open. Ostriker, Peebles, and Yahill (1974) introduced their paper on galaxy masses with the stunning sentence "There are reasons, increasing in number and quality, to believe that the masses of ordinary galaxies may have been underestimated by a factor of 10 or more." This work, plus that of Einasto, Kaasik, and Saar (1974) "The mass of galactic coronas exceeds that mass of populations of known stars by one order of magnitude, as do the effective dimensions," made use of arguments both observational and theoretical. They emphasized the evidence showing that masses of nearby giant spirals increase linearly with radius from 20 to 100 kpc and to as much as 500 kpc. These papers, coupled with an earlier paper (Ostriker & Peebles 1973) which demonstrated (with 150 to 500 mass points) that disk galaxies were "found

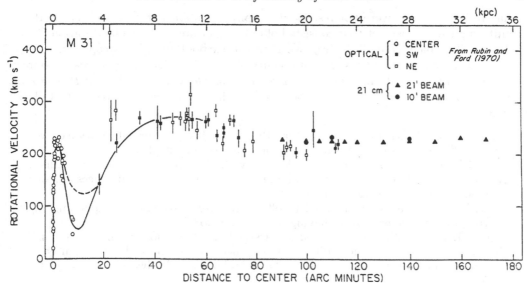

FIGURE 2. Rotation velocities in M31 as a function of distance from the nucleus. Optical Hα velocities come from Rubin and Ford (1970); 21-cm velocities come from Roberts and Whitehurst (1975).

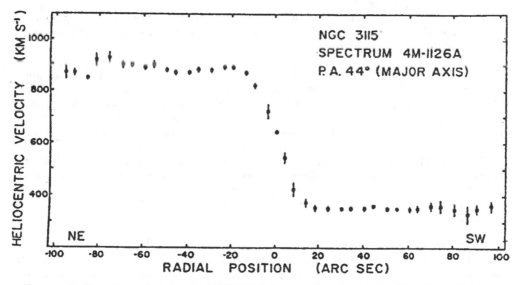

FIGURE 3. Rotation velocities for NGC 3115 from stellar absorption lines from Rubin, Peterson, and Ford (1976; @American Astronomical Society).

to be rapidly and grossly unstable to barlike modes," attracted considerable attention. It is surprising that it did not send scores of observers scurrying to their telescopes.

By 1975, the 21-cm studies by Mort Roberts and his collaborators produced (Roberts & Whitehurst 1975) a rotation curve for M31 (Fig. 2) that extended to 35 kpc (175′), almost 50% farther than the optical curve (Rubin & Ford 1970). At least for M31, there was no longer room for argument: the rotation curve was mathematically flat, over more than 50% of the detectable disk, much of it beyond the optical galaxy. For this best observed disk galaxy, astronomers knew that mass rose linearly with radius, and did not asymptotically approach a limit.

Flat rotation curves were not restricted to spirals. Rubin, Peterson, and Ford (1976) submitted an AAS abstract that consisted of only a plot and a title: The Rotation Curve of the E7/S0 Galaxy NGC 3115 (Fig. 3). It was a textbook plot: lat, steep, flat.

For a few astronomers' views of rotation curves in 1977, we have an impeccable source: a discussion at the Yale meeting, *The Evolution of Galaxies and Stellar Populations* (Tinsley & Larson 1977). I reproduce some of the discussion which followed the talk by Freeman.

> FREEMAN (to Krumm and Salpeter): Rumor has it that your flat rotation curves may be affected by sidelobe problems. Could you comment please?
>
> SALPETER: Sidelobe problems affect only the last point on a rotation curve. Even if the last point were removed, a large enough range of radial distances remains to demonstrate flat rotation curves.
>
> FREEMAN: Is anyone prepared to offer any alternative way of explaining the flat rotation curves aside from having a massive halo?
>
> M. ROBERTS: Why do you need a massive halo? Why not an unmassive disk? The mass goes up with the radius so it only need to be doubled. All you need is the most common type of star in the solar neighborhood, M dwarfs, to account for the M/L.
>
> The problem with the flat rotation curves is not peculiar to only one telescope, but is common to all the large telescopes that have been used. Any given case may be arguable, but overall there do seem to be flat rotation curves—although not ALL rotation curves are flat.
>
> L. SMITH: If it were true that 50% of stars formed were below 0.1 $M_\odot$ and if it were true that only MASSIVE star formation stops at $\approx$ 13 kpc galactic radius, would that solve the problem.
>
> M. ROBERTS: Yes.
>
> KING: Although I'm not particularly fond of massive halos, I can think of one argument that would favor an extended mass being in a halo rather than in a disk. This is based on the Westerbork observations of the high incidence of disks with twisted edges. If a twisted disk has much mass in its outer parts, it is very hard for it to maintain a clean twist without turning into a washboard. But it is much easier to maintain the twisted shape if the mass is in a halo instead.
>
> OSTRIKER: From observations by Hy Spinrad and myself, it does seem difficult to account for the mass with ordinary late M dwarfs, for they would give too much light. Of course, stars of even lower mass are possible. My other point is that if all the mass is in a flat, cold disk it is very likely to be unstable. But since the mass is invisible, it could be in a flat, hot disk.

As these exchanges indicate, astronomers were willing, in 1977, to accept that some rotation curves were flat. But at the same meeting, Rees (1977) focused his talk, Galaxy Formation, on the "implications of massive halos and 'missing mass' (which, if the participants in this conference are an unbiased sample, are seriously believed to exist...)." Yet not a single author referenced Zwicky's studies of dark matter in clusters of galaxies.

The next year, Bosma (1978) completed his thesis, observing and compiling 21-cm rotation curves for 25 galaxies (Fig. 4). All but a few had flat or almost flat curves. Only M81, M51, and M101 showed significant outer falling velocities; explanations in terms of tidal interactions would ultimately arise. Also in 1978, Kent Ford and I (Rubin & Ford 1978) published data for eight rotation curves of high luminosity spirals, and photos (Fig. 5) which showed their emission line spectra all strikingly flat to the eye. I think that Bosma's plots, plus the visual spectra, convinced many astronomers that rotation curves are flat. Not flat was the exception. But there were still non-believers. One eminent astronomer said to me "When you observe low luminosity galaxies, you'll find Keplerian falling rotation curves." Not so, of course. We know now that the lower

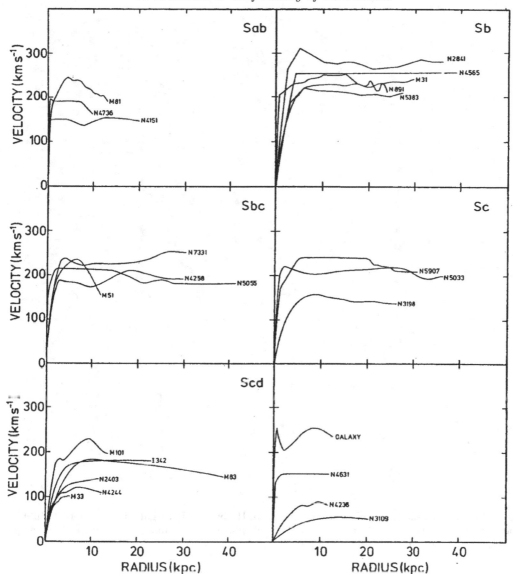

FIGURE 4. 21-cm rotation curves for 25 galaxies, from Bosma's (1978) thesis.

the luminosity, the fractionally more dark matter required. Kalnajs' (1983) insistence that dark matter is not required, at least for a few galaxies with spatially limited data, convinced a few astronomers that dark matter could be avoided. In retrospect, I think it is fair to say that many astronomers hoped that Kalnajs was right; dark matter was to be avoided, if at all possible.

In their review "The Kinematics of Spiral and Irregular Galaxies," van der Kruit and Allen (1978) waffled in their conclusion. "It is certainly true that more mass is waiting to be found beyond the last measured HI points on many rotation curves... However, the great increase (factors of 10 to 100) in masses favored by Einasto et al. (1974) and by Ostriker et al. (1974) involve estimates at much greater distances, from 200–500 kpc. ...There is no evidence in favor of such massive halos *within* the visible disks of galaxies...

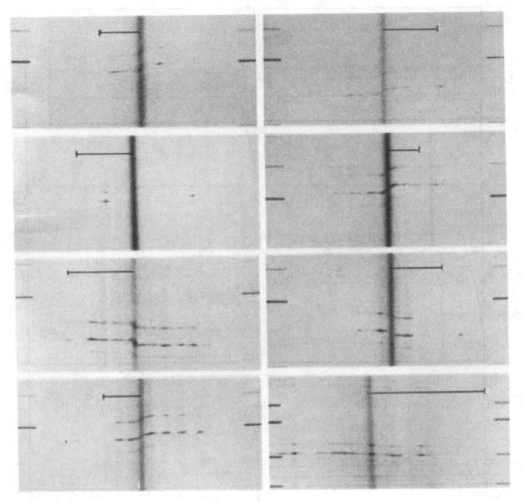

FIGURE 5. Spectra showing emission lines (dark) of Hα and [NII] for galaxies of different Hubble types, taken with the *Kitt Peak* 4-m spectrograph plus Carnegie image tube. Exposures range from 120 to 200 minutes (Rubin and Ford 1978).

We must conclude that the results from rotation curves are not inconsistent with the existence of extensive, massive halos around galaxies, although the prime evidence for them comes from studies of binary galaxies and outlying globular clusters (e.g. Turner and Ostriker 1977, Sargent 1977)."

Only one year later, a comprehensive review by Faber and Gallagher (1979) concluded more emphatically "After reviewing all the evidence, it is our opinion that the case for invisible mass in the Universe is very strong and getting stronger." Finally, a paper (since Hubble 1936) had united in print Zwicky's dark matter with flat rotation curves of galaxies.

Ostriker (1999) chose Zwicky's (1937c) *Astrophysical Journal* paper "On the Masses of Nebulae and Clusters of Nebulae" for reprinting in the *ApJ Centennial Issue* (Abt 1999). In a glowing discussion of Zwicky's paper, Ostriker wrote "Thus, we (Ostriker et al. 1994) took comfort in the fact that our estimates for the total mass were consistent with those reached by Zwicky decades earlier." Yet even this ground-breaking paper of

Ostriker et al. (1994) failed to reference Zwicky. However, these authors were in good company. A decidedly incomplete survey of papers in the 1960s and 1970s (but which does include most of the relevant papers referenced in the present paper) turned up not a single paper pre-Faber and Gallagher (1979) that referenced Zwicky.

Following the influential Faber and Gallagher (1979) review, it was "general belief" that rotation curves were flat. As more rotation curves accrued, dark matter became the accepted cause. But in as much as we have not yet succeeded in identifying the composition of dark matter, attributing flat rotation curves to 'dark matter' seems at times only a semantic construct.

5. Another approach: Gravitational lenses

As usual, Zwicky (1937a, 1937b, 1937c) was right. Gravitational lenses do offer a method for inferring the existence of dark matter in clusters of galaxies.

Early work concentrated on single galaxies as lenses. Just a few years after detecting the deflection of light by the sun during the 1919 total solar eclipse, Frost (1923, mentioned in Zwicky 1937b) suggested searching for the gravitational deflection of background sources by stars. A decade later, Zwicky (1937a, 1937b, 1937c) realized that the more massive galaxies would be far more important as lenses. Refsdal (1964) cites the literature through 1964, and Barnothy (1989) traces the detailed history preceding his suggestion (Barnothy 1965) that QSOs were lensed Seyfert galaxies.

Clusters of galaxies as gravitational lenses have a convoluted history. During tests of a SIT Vidicon, Jim Westphal (1973) observed the cluster Abell 370 with the 200-inch telescope. Images comparing the cluster photographed with the SIT Vidicon and an unaided photographic plate were published by *Science* magazine. In retrospect, both images are very noisy, but do show a gravitational arc, not mentioned. Hoag (1981) discussed a filament-like feature (the arc) in A370 at an AAS meeting; the source of the image is a *KPNO* 4-m prime focus plate.

A decade after the SIT Vidicon images, Butcher, Oemler, and Wells (1983) published a beautiful *Kitt Peak* 4-m prime focus image of A370 (Fig. 6), taken for their detailed study of galaxies in clusters. The very sharp and prominent arc is not referred to in the paper. When I sent an email request to Butcher and Oemler to reproduce the image in this paper, I gently warned them that is was included in papers which had been published, but had not noted the arc. Oemler replied (almost instantly) "Of course you may use it, and of course you may mention that we stared at that damn arc for endless hours without recognizing it—I always do."

Several years later, Lynds and Petrosian (1986) presented a paper "On the Giant Luminous Arcs in Clusters" at the AAS meeting. They noted their properties: spatially coherent, narrow arc-like shape of enormous lengths, whose radii of curvature point toward the central cD galaxies and the centers of gravity of the clusters. As in the previous work, a gravitational lens was not mentioned. A discussion with Lynds (private communication, 2000) suggests that he and Petrosian were considering various explanations, but were waiting to get a good spectrum of the arc.

Soucail and colleagues solved the puzzle, following a false start. Initially they (Soucail et al. 1987a) attributed the arcs "in" clusters to a characteristic of the cluster, perhaps a cooling flow." Katz (1987) and Milgrom (1987) thought them to be light echos from a beaming cluster source. Later, from initial spectra of poor quality, Soucail et al. (1987b) identified the arc in A370 as a gravitational lens. The next year, Soucail et al. (1988), in a paper received by *A&A* on November 17, 1987, showed a beautiful spectrum taken with an arc-shaped slit, which confirmed that the arc is the image of a galaxy at $z = 0.724$,

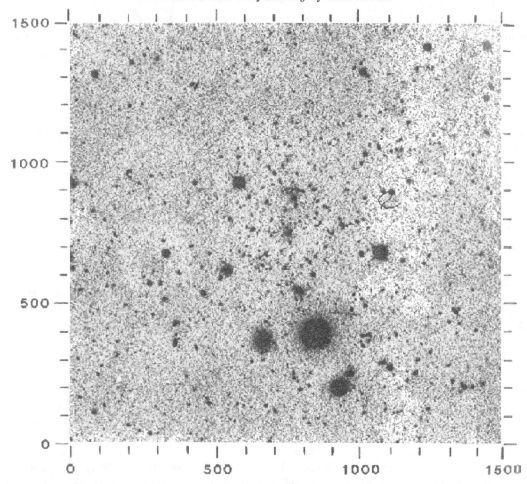

FIGURE 6. *Kitt Peak* 4-m prime focus image of Abell cluster A370, from Butcher, Oemler, and Well (1983). Note the arc, which was not mentioned in the paper.

gravitationally lensed by A370 at $z = 0.374$. It is true, as textbooks instructed, that lensing is optimal for a mass midway between the observer and the lensed object.

At about the same time in late 1987, Lynds and Petrosian (1988) submitted a late paper for the AAS meeting in Austin, Texas, January 1988. They too had obtained spectra and redshifts and announced the arc as the lensed image of a background galaxy. Their detailed paper (Lynds and Petrosian 1989) gives a slight history leading up to this conclusion.

6. Conclusion

By the end of the decade of the 70s, astronomers generally agreed that rotation velocities did not show a Keplerian decline toward the edge of the optical disk. An extended spheroidal distribution of matter, roughly centered on the galaxy center, more extended than the optical galaxy, and more massive that the luminous galaxy, was the generally adopted solution. Today, although we think we know the quantity of luminous matter in the universe, and can put a limit on the amount of dark matter in the universe, we have made little progress in deducing the composition of the dark matter.

For a few scientists (Finzi 1963; Bekenstein & Milgrom 1984, Milgrom 1983; Sanders 1990, McGaugh and de Blok 1998), flat rotation curves have been a sufficient reason to investigate alternatives to Newtonian gravitational theory. Rather than discuss these alternatives which are generally as valid as dark matter in fitting the observations (although gravitational lensing might stress the theories), I will close with a few historical comments relating to scales in nature.

At the close of the 19th century, physicists were discovering phenomena that did not fit into the science they had grown up with. Late in life, in discussing the early years of quantum mechanics, Heisenberg (1961) wrote "Although the previous laws of nature, e.g., Newtonian mechanics, contained so-called constants, the constants referred to the properties of objects... On the other hand, Planck's action quantum, which is the characteristic constant in his law of radiation, does not represent a property of objects, but a property of nature. It establishes a scale in nature."

Heisenberg continues "While the laws of former physics, e.g., Newtonian mechanics, should basically be valid for all orders of magnitude (the movement of the moon around the earth should obey the same laws as the fall of the apple from the tree or the deviation of an alpha particle that grazes the nucleus of an atom), Planck's law of radiation shows for the first time that there are scales in nature, that phenomena in different ranges of magnitude are not necessarily of the same type." It is interesting that Heisenberg's large scale example does not extend beyond the solar system.

Feynman, in "The Character of Physical Law" (1965) starts with the solar system, and the problem with the apparent motion of Mercury which "was shown by Einstein that Newton's Laws were slightly off and then they had to be modified. The question is, how far does this law extend? Does it extend outside the solar system?" After discussing a binary star system and a globular cluster, he shows a "typical galaxy, and it is clear once again that this thing is held together by some force, and the only candidate that is reasonable is gravitation. When we get to this size we have no way of checking the inverse square law, but there seem to be no doubt that in these great agglomerations of stars... gravity is extending even over these distances."

He ends by naming a characteristic that gravity shares with other physical laws. "It is mathematical, ...it is not exact; Einstein had to modify it, and we know that it is not quite right yet, because we have still to put the quantum theory in. That is the same with all our other laws—they are not exact. There is always an edge of mystery, a place where we have some fiddling around to do yet. This may or may not be a property of Nature, but it is certainly common to all the laws as we know them today. It may be only a lack of knowledge."

Only the future will tell us what the dark matter is, or whether our lack of knowledge of gravitation on the largest scales has fooled us. It will be exciting to follow the path that leads us from this edge of mystery to answer the question, "What is dark matter?"

REFERENCES

ABT, H. A. 1999 *American Astronomical Society Centennial Issue* **525**, Number 1C, Part 3.
AMBARTSUMIAN, V. A. 1961 *AJ* **66**, 536.
BABCOCK, H. W. 1939 *Lick Obs. Bull.* **19**, 41.
BARNOTHY, J. M. 1965 *AJ* **70**, 666.
BARNOTHY, J. M. 1989, in *Gravitational Lenses: Proceedings of a Conference Held in Honor of Bernard F. Burke's 60th Birthday* (eds. J. M. Moran, J. N. Hewitt, & K. Y. Lo). p. 23. Springer.
BEKENSTEIN, J. & MILGROM, M. 1984 *ApJ* **286**, 7.
BOSMA, A. 1978 *Ph.D. Thesis*, University of Groningen.

BURBIDGE, E. M., BURBIDGE, G. R., & PRENDERGAST, K. H. 1962 *ApJ* **136**, 128.

BUTCHER, H., OEMLER, A., & WELLS, D. C. 1983 *ApJS* **52**, 183.

CLERKE, AGNES MARY 1885, in *A Popular History of Astronomy during the Nineteenth Century*, Adam and Charles Black. p. 41

DE VAUCOULEURS, G. 1959, in *Handbuch der Physics, Astrophysik IV* **53**, 310.

EINASTO, J., KAASIK, A., & SAAR, E. 1974 *Nature* **250**, 309.

FABER, S. M. & GALLAGHER, J. S. 1979 *ARA&A* **17**, 135.

FEYNMAN, R. 1965, in *The Character of Physical Law*. MIT Press.

FINZI, A. 1963 *MNRAS* **127**, 21.

FREEMAN, K. C. 1970 *ApJ* **160**, 811.

HEISENBERG, W., BORN, M., SCHRODINGER, E., & AUGER, P. 1961 *On Modern Physics*. p. 7. Paolo Boringhieri Editore.

HOAG, A. 1981 *BAAS* **13**, 799.

HUBBLE, E. 1936, in *The Realm of the Nebulae*. p. 180. Yale Univ. Press.

JEANS, J. H. 1919, in *Problems of Cosmogony and Stellar Dynamics*. p. 239. Cambridge Univ. Press.

JEANS, J. H. 1922 *MNRAS* **82**, 130.

KAHN, F. D. & WOLTJER, L. 1959 *ApJ* **130**, 705.

KALNAJS, A. J. 1983, in *Internal Kinematics and Dynamics of Galaxies*, IAU Symposium 100 (ed. A. Athanassoula). p. 87. Reidel.

KAPTEYN, J. C. 1922 *ApJ* **55**, 302.

KATZ, J. I. 1987 *A&A* **182**, L19.

KRUMM, N. & SALPETER, E. E. 1976 *ApJ* **208**, L7.

LYNDS, R. & PETROSIAN, V. 1986 *BAAS* **18**, 1014.

LYNDS, R. & PETROSIAN, V. 1988 *BAAS* **20**, 644.

LYNDS, R. & PETROSIAN, V. 1989 *ApJ* **336**, 1.

MCGAUGH, S. S. & DE BLOK, W. J. G. 1998 *ApJ* **499**, 66.

MAYALL, N. U. 1951, in *The Structure of the Galaxy*, p. 19. University of Michigan Press.

MILGROM, M. 1983 *ApJ* **270**, 371.

MILGROM, M. 1987 *A&A* **182**, L21.

MITCHELL, REV. JOHN 1784 *Phil. Trans. Royal. Soc. London*.

OORT, J. H. 1932 *Bull. Ast. Neth.* **6**, 249.

OORT, J. H. 1940 *ApJ* **91**, 273.

OSTRIKER, J. P. 1999, in *American Astronomical Society Centennial Issue* (ed. H. A. Abt) **525**, Number 1C, Part 3, p. 297.

OSTRIKER, J. P. & PEEBLES, P. J. E. 1973 *ApJ* **186**, 467.

OSTRIKER, J. P., PEEBLES, P. J. E., & YAHILL, A. 1974 *ApJ* **193**, L1.

REES, M. J. 1977, in *The Evolution of Galaxies and Stellar Populations* (eds. B. M. Tinsley & R. B. Larson). p. 339. Yale University Observatory.

REFSDAL, S. 1964 *MNRAS* **128**, 23.

ROBERTS, M. A. & ROTS, A. H. 1973 *A&A* **26**, 483.

ROBERTS, M. S. & WHITEHURST, R. N. 1975 *ApJ* **201**, 327.

ROGSTAD, D. H. & SHOSTAK, G. S. 1972 *ApJ* **176**, 315.

RUBIN, V. C., BURLEY, J., KIASATPOOR, A., KLOCK, B., PEASE, G., RUTSCHEIDT, E., & SMITH, C. 1962 *AJ* **67**, 491 [I had to fight with the AJ editor to get the students named joint authors].

RUBIN, V. C. 1965 *ApJ* **142**, 934.

RUBIN, V. C., BURBIDGE, E. M., BURBIDGE, G. R., & CRAMPIN, D. J. 1965 *ApJ* **141**, 759.

RUBIN, V. C. & FORD, W. K. JR. 1970 *ApJ* **159**, 379.

RUBIN, V. C. & FORD, W. K. JR. 1978 *ApJ* **225**, L107.

RUBIN, V. C., PETERSON, C. J., & FORD, W. K. JR. 1976 *BAAS* **8**, 297.

SANDERS, R. H. 1990 *A&AR* **2**, 1.

SARGENT, W. L. W. 1977, in *The Evolution of Galaxies and Stellar Populations* (eds. B. M. Tinsley & R. B. Larson). p. 427. Yale University Observatory.

SCHWARZSCHILD, M. 1954 *AJ* **59**, 273.

SMITH, S. 1936 *ApJ* **83**, 23.

SOUCAIL, G., FORT, B., MELLIER, Y., & PICAL, J. P. 1987a *A&A* **172**, L14.

SOUCAIL, G., MELLIER, Y., FORT, B., HAMMER, F., & MATHEZ, G. 1987b *A&A* **184**, L7.

SOUCAIL, G., MELLIER, Y., FORT, B., MATHEZ, G., AND CAILLOUX, M. 1988 *A&A* **191**, L19.

TINSLEY, B. M. & LARSON, R. B. (eds.) 1977 *The Evolution of Galaxies and Stellar Populations*, p. 174. Yale University Observatory.

TRIMBLE, V. 1987 *ARA&A* **25**, 425.

TRIMBLE, V. 1995, in *Dark Matter* (eds. S. S. Holt & C. L. Bennett). p. 57. AIP.

TURNER, E. L. & OSTRIKER, J. P. 1977 *ApJ* **217**, 24.

VAN DE HULST, H. C., RAIMOND, E., & VAN WOERDEN, H. 1957 *Bull. Astro. Neth.* **14**, 1.

VAN DEN BERGH, S. 1961 *AJ* **66**, 566.

VAN DEN BERGH, S. 1999 *PASP* **111**, 657.

VAN DER KRUIT, P. C. & ALLEN, R. S. 1987 *ARA&A* **16**, 103.

VOLDERS, L. 1959 *BAN* **1**, 323.

WHITMAN, WALT 1855, in *Song of Myself*.

WESTPHAL, J. A. 1973 *Science* **181**, 930.

ZWICKY, F. 1933 *Helv. Phys. Acta.* **6**, 110.

ZWICKY, F. 1937a *Phys. Rev.* **51**, 290.

ZWICKY, F. 1937b *Phys. Rev.* **51**, 679.

ZWICKY, F. 1937c *ApJ* **86**, 217.

Microlensing towards the Magellanic Clouds: Nature of the lenses and implications on dark matter

By KAILASH C. SAHU

Space Telescope Science Institute, 3700 San Martin Drive, Baltimore, MD 21218, USA;
ksahu@stsci.edu

A close scrutiny of the microlensing results towards the Magellanic clouds reveals that the stars within the Magellanic clouds are major contributors as lenses, and the contribution of MACHOs to dark matter is 0 to 5%. The principal results which lead to this conclusion are the following:

(i) Out of the ~17 events detected so far towards the Magellanic Clouds, the lens location has been securely determined for one binary-lens event through its caustic-crossing timescale. In this case, the lens was found to be within the Magellanic Clouds. Although less certain, lens locations have been determined for three other events and in each of these three events, the lens is most likely within the Magellanic clouds.

(ii) If most of the lenses are MACHOs in the Galactic halo, the timescales would imply that the MACHOs in the direction of the LMC have masses of the order of $0.5\ M_\odot$, and the MACHOs in the direction of the SMC have masses of the order of 2 to $3\ M_\odot$. This is inconsistent with even the most flattened model of the Galaxy. If, on the other hand, they are caused by stars within the Magellanic Clouds, the masses of the stars are of the order of $0.2\ M_\odot$ for both the LMC as well as the SMC.

(iii) If 50% of the lenses are in binary systems similar to the stars in the solar neighborhood, ~10% of the events are expected to show binary characteristics. The fact that the two observed binary events are caused by lenses within the Magellanic Clouds would imply that there should be a total of about 20 events caused by lenses within the Magellanic Clouds. This implies that most of the microlensing events observed so far are probably caused by stars within the Magellanic Clouds.

(iv) If the microlensing events are caused by MACHOs of $0.5\ M_\odot$, as claimed from the LMC observations, about 15 events should have been detected by now towards the SMC, with timescales of ~40 days. The fact that no event has been detected towards the SMC caused by MACHOs (both the events detected towards the SMC have been shown to be due to self-lensing) places severe constraints on the MACHO contribution and suggests that the contribution of MACHOs to dark matter is consistent with zero, with an upper limit of 5%.

---

1. Introduction

Most spiral galaxies are known to have flat rotation curves (e.g. Begeman et al. 1991). In the outer parts of the galaxies in particular, the visible matter falls far short of what is required to explain the flat rotation curves. In our own Galaxy, the rotation curve is observed to be flat up to at least 16 kpc from the center (e.g. Fich et al. 1989). The scenarios proposed to explain the flat rotation curves are either departure from Newtonian dynamics in large scale, or presence of dark matter (for a review see Sanders 1990). In the latter and more conventional scenario, the rotation curves would imply that a significant part of the mass of the galaxies, including our own, resides in the halo in some form of dark matter. The nature of this dark matter has been hypothesized to be in either of the two forms: MACHOs (i.e. massive compact halo objects which is a collective term for 'Jupiters,' brown dwarfs, red dwarfs, white dwarfs, neutron stars or black holes); or elementary particles such as massive neutrinos, axions or WIMPs (weakly interacting massive particles such as photinos, etc.). One important distinguishing factor

14

of the MACHOs is that they can have detectable gravitational effects and can cause "gravitational microlensing" of background stars.

Einstein (1936) was the first to point out that a star can act as a gravitational lens for another background star, if the two are sufficiently close to each other in the line of sight. Given the observational capabilities of that time, Einstein had however considered this to be a purely theoretical exercise since there was "no hope of observing such a phenomenon directly."

Paczyński (1986) worked out the probability of such microlensing events by MACHOs and showed that if the halo of our Galaxy is made up of MACHOs, the probability of finding any given star being lensed by them is 5×10^{-7}, independent of their mass distribution. He suggested an experiment to look for such events using the LMC stars. Such an experiment was taken up by two groups who reported their first results in 1993 (Alcock et al. 1993; Aubourg et al. 1993). In six years' monitoring of millions of stars towards the Magellanic Clouds, so far about 17 events have been detected towards the Large Magellanic Cloud (LMC) and two events towards the Small Magellanic Cloud (SMC).

The observed microlensing optical depth as derived from these observations is $\sim 1 \times 10^{-7}$. This is about a factor of 5 lower than what would be expected if the dark matter is entirely made up of MACHOs. To complicate the issue further, the simple microlensing light curves cannot tell us the location of the lenses because of a degeneracy between the distance, mass and the proper motion of the lens. So the lenses can be stars in the Galactic disk (Gould 1994; Evans et al. 2001; Gates & Gyuk 2001), MACHOs in the Galactic halo (Alcock et al. 2000a), or stars within the LMC itself (Sahu 1994a,b; Evans & Eamonn 2000). As a result, the interpretations of these lensing events, and their locations in particular, have remained uncertain.

2. Overview of the problem

The contribution of known stars to the microlensing optical depth had not been calculated until after the first microlensing events were reported towards the LMC. Such calculations revealed that the microlensing optical depth due to the known stars within the LMC is close to the observed optical depth, particularly in the region of the bar where the microlensing event was detected (Sahu 1994a,b; Wu 1994). This was used to argue that the stars within the Magellanic Clouds must play a significant role as lenses.

Lensing by stars within the Magellanic clouds is now commonly referred to as "self-lensing." It has been known for a long time (e.g. Schneider et al. 1992), and it was re-derived by Gould (1995) that the self-lensing microlensing optical depth can be expressed as a function of the velocity dispersion

$$\tau = 2 \times sec^2(\theta)v^2/c^2 \ , \tag{2.1}$$

where θ is the inclination of the disk of the LMC (the best estimated value of which is 34.7 ± 6 deg, van der Marel and Cioni, 2001). Since the velocity dispersion of the stars in the disk of the LMC is about 20 km s^{-1}, the above equation would imply that the self-lensing optical depth is about 2×10^{-8}, which is too low to explain the observed microlensing events.

However, there are some problems in such an approach. First, eq. (2.1) can be applied only for a virialized system. The Magellanic Clouds are not relaxed, and far from virialized systems: they are dynamically disturbed, they show tidal-tail structures, and the velocity dispersions are different for low and high-mass objects (for a review, see Westerlund 1997). Second, the lenses are low-mass objects for which the velocity dispersions are unknown.

Third, the velocity dispersions in the region of the bar—where most of the events have been found so far—is also unknown.

A cautionary note seems appropriate here. A similar story had repeated towards the Galactic bulge. The velocity dispersion of the Galactic bulge stars would imply a microlensing optical depth which is a factor of at least three smaller than the observed optical depth. In the case of the Galactic bulge, the re-discovery of the Galactic bar, with an inclination of 15° to the line of sight helped in solving the puzzle of the discrepancy between the observed and the expected microlensing optical depth (Kiraga & Paczyński, 1994; Paczyński et al. 1995).

In the meantime, several papers have appeared with detailed calculations of the optical depth taking into account the extra effects such as 'diffusion' of small-mass objects, which causes the velocity dispersion of less massive objects to be larger (e.g. Salati et al. 1999).

The full range of the calculated optical depth due to the stars within the LMC, as published in the literature so far, is 1×10^{-8} to 2×10^{-7} (e.g. Gyuk et al. 2000, Graff 2001, Salati et al. 1999, Aubourg et al. 1999). The high end of this self-lensing optical depth would imply that all the observed events are caused by stars within the LMC and that the contribution of MACHOs to the dark matter is negligible. The low-end of the self-lensing optical depth would imply that none of the observed events are caused by stars within the Magellanic clouds, and hence the events are caused most likely by MACHOs in the halo. In the extreme case where all the events are caused by MACHOs in the Galactic halo, the implied contribution of MACHOs to dark matter is about 25% (Alcock et al. 2000a; Afonso et al. 2002).

Clearly, it is important to know the exact location and nature of the lenses, which can shed some light on the whether MACHOs can account for the long-sought dark matter in the halo. We obviously need some observational tests to guide us in resolving this puzzle. The rest of the paper mainly deals with such tests and the current status of the results from these tests.

It is worth pointing out here that, in the self-lensing hypothesis, it is implicit that the sources and the lenses reside within the LMC. The sources cannot be considered as background sources since the derivation of the observed microlensing optical depth assumes that all the monitored stars which are *within* the Magellanic clouds contribute to the microlensing optical depth.

3. Observational tests

Fortunately, there are several observational tests which can be used to infer the nature of the lenses, including some tests which can be used for direct determinations of their locations. The observational tests that I will discuss here, some of which have already provided clear results, are the following:

    (i) Determination of lens location through "caustic-crossing" timescale,

    (ii) Other direct determinations of lens locations,

    (iii) Frequency of observed binary-lens events,

    (iv) Timescales of SMC vs. LMC events, and

    (v) Spatial distribution of the microlensing events.

3.1. *"Caustic-crossing" timescale*

Clearly, the best test would be to measure the location of the lens directly. In a few special cases, such a direct determination of the lens location is indeed possible. One such special case is when the lens is a binary, and the source crosses the caustics produced by the binary lens. Since a caustic is essentially a straight line in space, the time taken by the

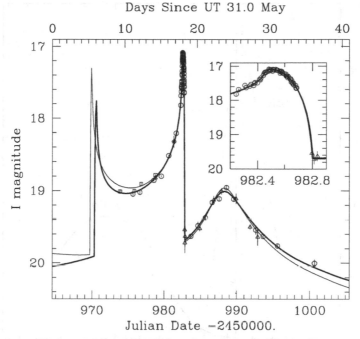

FIGURE 1. Light curve of the PLANET data for MACHO-98-SMC-1. Shown are the data from the SAAO 1 m (*circles*), the CTIO 0.9 m (*squares*), the CTIO-Yale 1 m (*triangles*), and the Canopus 1 m (*asterisks*). The inset covers about 0.6 days, corresponding to less than one tick mark on the main figure. (Taken from Albrow et al. 1999)

caustic to cross the source provides a direct measure of the proper motion of the lens projected onto the source plane. If the lens is in the halo at a distance of ~10 kpc, then the expected proper motion of the lens is about 200 km s^{-1} which, projected onto the source plane, is about 1000 km s^{-1}. So the time for the caustic to cross the source would be of the order of half an hour. If, on the other hand, the lens is within the Magellanic Clouds, the expected proper motion is about 50 km s^{-1}, and hence the caustic crossing time is expected to be of the order of 10 hours. Thus, monitoring a caustic crossing provides a powerful method to determine the location of the lens.

3.1.1. *MACHO 98-SMC-1*

Such an opportunity presented itself after the first binary-lens event, MACHO 98-SMC-1, was discovered by the MACHO collaboration in 1998 (Alcock et al. 1999). After the first caustic crossing was reported, the PLANET collaboration, with its 24-hour access to telescopes around the world, predicted a second caustic crossing and began monitoring this event, with a particular interest in fully sampling the second caustic crossing (Albrow et al. 1999). The time for the caustic to cross the source was found to be 8.5 hours (Fig. 1), which demonstrated that the lens is within the SMC (Fig. 2). The result was further confirmed by the EROS, MACHO and MPS collaborations, who also concluded that the lens is most likely within the SMC (Afonso et al. 1998, 2000, Alcock et al. 1999, Rhie et al. 1999).

3.1.2. *MACHO-LMC-9*

The second binary-lens event, MACHO-LMC-9, was discovered by the MACHO collaboration towards the LMC, for which the observations are not as dense (Alcock et al.

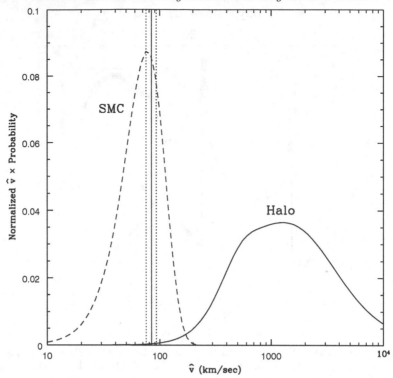

FIGURE 2. Predicted velocity distributions (at the source plane) for halo and SMC lenses. The measured value of the velocity and error bars are shown as vertical lines. The observed value is inconsistent with the halo hypothesis. The lens and source could possess normal SMC kinematics. (Taken from Alcock et al. 1999)

1997a). The caustic crossing timescale of the event also suggests that the lens is most likely within the LMC (Fig. 3), although the sparse sampling of this event quality has led to some doubt as to whether this is a microlensing event at all (Alcock et al. 2000a,b).

3.2. *Other determinations of lens locations*

3.2.1. *MACHO 97-SMC-1*

97-SMC-1 is the first microlensing event observed towards the SMC, which has a large timescale of 220 days, in contrast to the average timescale of 40 days as observed for the more than dozen microlensing events towards the LMC (Alcock et al. 1997b).

If the lens is a MACHO in the Galactic halo, the timescale of the event would suggest that the mass of the lens is about 3 $M_\odot$. Clearly, this cannot be a normal star of 3 $M_\odot$ since it would be too bright and would be easily detectable. Indeed, a spectrum of the source obtained by Sahu & Sahu (1999) shows the spectrum to be a pure B-type spectrum as expected from a source within the SMC, without any contribution from the lens (Fig. 4). This would suggest that the lens is either a black hole in the Galactic halo, or a small mass object within the SMC.

A black hole in the Galactic halo is an unsatisfactory explanation since that would imply that the MACHOs in the direction of the LMC are predominantly 0.5 solar mass objects and the MACHOs in the direction of the SMC are predominantly 3 solar mass objects. The locations of the LMC and the SMC too close to permit such a mass segregation in these two directions even in the most extreme flattened model of the Galaxy.

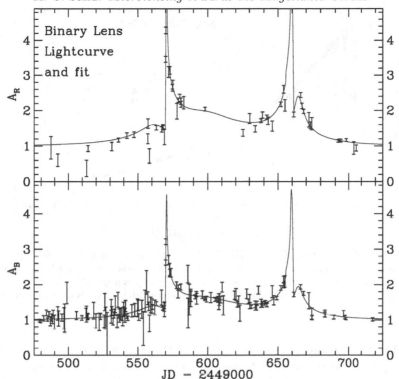

FIGURE 3. Light curve of LMC-9 in the around the peak of the event, and the best-fit binary lens light curve. The most likely lens location is within the LMC. (Taken from Alcock et al. 1997a.)

This leaves us with the only other alternative that the lens must be a small mass object within the SMC.

Furthermore, as explained in the next section, this long timescale of the SMC events is precisely what one would expect if the lenses are within the Magellanic Clouds, which further supports the view that the lens is most likely within the SMC.

3.2.2. *MACHO 96-LMC-2*

Lens locations have been estimated for another event MACHO 96-LMC-2, where the source is a binary (Alcock et al. 2001a). In this case too, the lens is most likely within the LMC.

Thus, there are four cases so far for which we know the locations of the lenses. In all the cases, without exception, the most likely locations of the lenses are within the Magellanic Clouds.

It is worth noting here that the lens location has been recently determined for an event caused by a star in the local Galactic disk (Alcock et al. 2001b). The events caused by nearby lenses are expected to exhibit "parallax" effects which are easier to detect. The fact that only one event caused by a star in the local Galactic disk has been detected is consistent with the expected frequency of nearby stars acting as lenses.

3.3. *Frequency of observed binary-lens events*

For the sake of argument, let us start with the assumption that all the events except the two binary events for which we know that the lenses are within the Magellanic clouds are caused by MACHOs in the halo. This would be quite extra-ordinary since this would

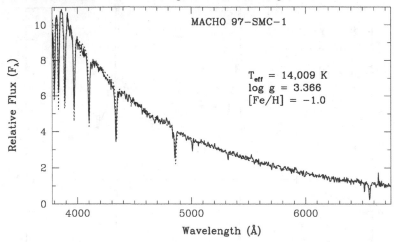

FIGURE 4. Observed spectrum of MACHO 97-SMC-1 taken on 30th May 1997, just after the source had crossed the Einstein ring radius of the lens (solid curve), along with the best fit stellar model spectrum (dashed line). The contribution from the lens is negligible which, combined with the microlensing time scale, implies that the lens must be within the SMC. (Taken from Sahu & Sahu 1998).

mean that the only two events that are caused by the stars within the Magellanic clouds not only happen to be binaries, but they also show caustic crossing structure in their light curves. If one event caused by a star in the Magellanic clouds is observed, the probability that the lens would be a binary is about 50% (assuming 50% of the lenses are in binary systems similar to the stars in the solar neighborhood). Since only 20% of the binary-lens events are expected to *show* caustic-crossing structures (which is borne out by the observations towards the Galactic bulge as well), the probability that it would *show* a caustic-crossing structure is 10%. The probability that a second event caused by a Magellanic cloud star would again be a binary and show caustic structure is again 10%. The probability that two successive events caused by Magellanic cloud stars will both be binaries and show caustic crossing structures is thus 1%, and hence extremely unlikely. We can follow the same argument in reverse and ask the following question: if two events which show binary characteristics are within the Magellanic clouds, how many of the observed events should be caused by stars within the Magellanic clouds? Since 10% of the events are expected to show binary characteristics, one would expect about 20 events to be caused by stars within the Magellanic clouds. This is more or less the number of events observed, which suggests that most of the lenses must be within the Magellanic clouds.

3.4. *Timescales of SMC vs. LMC events*

The relative timescales of the LMC and the SMC events provide further indications on the locations of the lenses. If the lenses are MACHOs in the Galactic halo, the characteristics of the lenses towards the LMC and the SMC are expected to be the same and hence both the LMC and the SMC events are expected to have similar timescales. On the other hand, if the lenses are within the Magellanic Clouds, the expected timescales of the LMC and the SMC events are expected to be very different. The LMC has a depth of less than one kpc in a typical line of sight whereas the SMC as a much larger depth of about five kpc along a typical line of sight. So the typical distance between the lens and the source is much smaller in the case of LMC than in the case of the SMC. As a result, the size

Line of Sight	No. of Events	M_{lens} (halo) ($M_\odot$)	M_{lens} (LMC/SMC) ($M_\odot$)	M_{lens} (local disk) ($M_\odot$)
LMC	~17	~0.5	~0.2	~0.5
SMC	~ 2	>2	~0.2	~2

TABLE 1. Lens masses for different scenarios

of the Einstein ring (which scales as the square root of the distance between the source and the lens) is much smaller for the LMC events than for the SMC events. Since the velocity dispersions of the stars within the LMC and the SMC are similar, the timescales of the SMC events are expected to be much longer than the LMC events, if the lenses are within the Magellanic Clouds.

Two events have been observed towards the SMC so far which have timescales of 75 and 125 days. This would correspond to lens masses of 2 to 3 solar masses if the lenses are in the halo, and about 0.2 to 0.3 solar masses if the lenses are within the Magellanic clouds. For the ~17 events observed towards the LMC, the timescales are much shorter, which would correspond to lens masses of ~0.5 solar masses if the lenses are in the halo, and ~0.2 solar masses if the lenses are within the LMC. This is shown it tabular form in Table 1. If the events are predominantly caused by MACHOs in the halo, one faces the unrealistic consequence that the MACHOs in the direction of the LMC and the SMC are of very different mass. Self-lensing is the only scenario which gives consistent masses for both the LMC and the SMC events. The longer durations of the SMC events are a natural consequence of the self-lensing hypothesis.

3.5. *Spatial distribution of the events*

Finally, the spatial distribution of the events provide some extra insight into the nature of the lenses. As first pointed out by Sahu (1994 a,b), the events should be more concentrated towards the bar of the LMC if the lensing is caused by the stars within the LMC. If the events are caused by MACHOs then the events (for a given number of monitored stars) should be uniformly distributed over the whole of the LMC. Unfortunately, all the analyses by the MACHO group so far have been confined to the region around the bar of the LMC. The EROS group, however, have mostly monitored in the region outside the bar, and have not detected any event. From their non-detection of microlensing events, they derive a microlensing optical depth which is much smaller than the optical depth determined by the MACHO group. This is consistent with the fact that stars within the Magellanic clouds play a dominant role as lenses.

Monitoring some regions far from the bar of the LMC to look for microlensing events would provide a clear test on whether the lensing events are caused by MACHOs or stars within the LMC (e.g. Stubbs 1999).

4. Implications on dark matter

The most significant insight into the contribution of dark matter, in my view, comes from a comparative study of the number of observed events towards the LMC and the SMC, and their timescales. So far, two microlensing events have been detected towards the SMC with time scales of 75 and 125 days. From these two events, the optical depth towards the SMC has been estimated to be $\sim 2 \times 10^{-7}$, which is about the same as the optical depth towards the LMC (Alcock et al. 1999). In terms of optical depth, these two events are equivalent to about 15 events with timescales of ~40 days (optical depth scales as the square of the timescale). Thus, if the microlensing events are caused by 0.5 solar mass MACHOs as claimed from the LMC observations, about 15 such events should have been detected towards the SMC by now. Furthermore, as discussed earlier, both of the observed SMC events have been shown to be caused by stars within the SMC. Thus no microlensing event caused by a MACHO has been detected towards the SMC, whereas 15 such events should have been detected if the contribution of MACHOs to dark matter is indeed 20% as claimed from the detection of ~20 microlensing events towards the LMC. This leads us to the inevitable conclusion that the contribution of MACHOs to dark matter is less than 2%, with a strong upper limit of 5%.

5. Conclusions

Results from several observational tests are already available which can be used to clearly distinguish between different scenarios for the nature of the lenses. I have argued that the results obtained so far point to the fact that the observed microlensing events towards the Magellanic clouds are predominantly caused by the stars within the Magellanic Clouds. Consequently, the contribution of MACHOs to the dark matter is consistent with zero, with a strong upper limit of 5%. However, some estimates suggest that the known stars within the Magellanic Clouds fall short by a factor of a few to explain the observed microlensing events. The contribution to the microlensing optical depth by the known stars is, however, very uncertain, and further work would be needed to resolve this issue.

REFERENCES

AFONSO, C., ET AL. 1998 *A&A* **337**, L17.

AFONSO, C., ET AL. 2000 *ApJ* **532**, 340.

AFONSO, C., ET AL. (EROS COLLABORATION) 2002 *A&A*, **400**, 951; astro-ph/0212176.

ALBROW, M., ET AL. 1999 *ApJ* **512**, 672.

ALCOCK, C., ET AL. 1993 *Nature* **365**, 621.

ALCOCK, C., ET AL. 1997a *ApJ* **491**, L11.

ALCOCK, C., ET AL. 1997b *ApJ* **486**, 697.

ALCOCK, C., ET AL. 1999 *ApJ* **518**, 44.

ALCOCK, C., ET AL. 2000a *ApJ* **542**, 281.

ALCOCK, C., ET AL. 2000b *ApJ* **541**, 270.

ALCOCK, C., ET AL. 2001a *ApJ* **552**, 259.

ALCOCK, C., ET AL. 2001b *Nature* **414**, 617.

AUBOURG, E., ET AL. 1993 *Nature* **365**, 623.

AUBOURG, E., ET AL. 1999 *A&A* **347**, 850.

BEGEMAN, K. G., BROEILS, A. H., & SANDERS, R. H. 1991 *MNRAS* **249**, 523.

EINSTEIN, A. 1936 *Science* **84**, 506.

EVANS, E. N. & EAMONN, K. 2000 *ApJ* **529**, 917.

EVANS, N. W., GYUK, G., TURNER, M. S., & BINNEY, J. 1998 *ApJ* **501**, L45.

FICH, M., BLITZ, L., & STARK, A. 1989 *ApJ* **342**, 272.

GATES, E. L. & GYUK, G. 2001 *ApJ* **547**, 786.

GOULD, A. 1994 *ApJ* **421**, L71.

GOULD, A. 1995 *ApJ* **441**, 77.

GRAFF, D. 2001. In *Microlensing 2000: A New Era of Microlensing Astrophysics* (eds. J. W. Menzies & Penny D. Sackett). ASP Conference Proceedings, vol. 239, p. 73. ASP.

GYUK, G., DALAL, N., & GRIEST, K. 2000 *ApJ* **535**, 90.

KIRAGA, M. & PACYŃSKI, B. 1994 *ApJ* **430**, L101.

PACZYŃSKI, B. 1986 *ApJ* **304**, 1.

PACZYŃSKI, B., STANEK, K. Z., UDALSKI, A., SZYMANSKI, M., KALUZNY, J., KUBIAK, M., MATEO, M., & KRZEMINSKI, W. 1994 *ApJ* **435**, L113.

PALANQUE-DELABROUILLE, N., ET AL. (EROS COLLABORATION) 1997 *A&A* **332**, 1.

PALANQUE-DELABROUILLE, N. 2001 *New Astron. Rev.* **45**, 395.

RHIE, S. H., ET AL. 1999 *ApJ* **522**, 1037.

SAHU, K. C. 1994a *Nature* **370**, 275.

SAHU, K. C. 1994b *PASP* **106**, 942.

SAHU, K. C. & SAHU, M. S. 1998 *ApJ* **508**, 147.

SALATI, P., ET AL. 1999 *A&A* **350**, L57.

SANDERS, R. H. 1990 *A&A* **2**, 1.

SCHNEIDER, P., EHLERS, J., & FALCO, E. E. 1992 *Gravitational Lensing*. Springer-Verlag.

STUBBS, C. 1999. In *The Third Stromlo Symposium: The Galactic Halo* (eds. B. K. Gibson, T. S. Axelrod, & M. E. Putman). ASP Conference Series, vol. 165, p. 503. ASP.

VAN DER MAREL, R. & CIONI, M. L. 2001 *AJ* **122**, 1807.

WESTERLUND, B. E. 1997 *The Magellanic Clouds*. Cambridge University Press.

WU, X-P. 1994 *ApJ* **435**, 66.

Searching for the Galactic dark matter

By HARVEY B. RICHER

Department of Physics & Astronomy, University of British Columbia, 6224 Agricultural Road, Vancouver, B.C., V6T 1Z1, Canada; richer@astro.ubc.ca

A straightforward interpretation of the MACHO microlensing results in the direction of the Magellanic Clouds suggests that an important fraction of the baryonic dark matter component of our Galaxy is in the form of old white dwarfs. If correct, this has serious implications for the early generations of stars that formed in the Universe and also on the manner in which galaxies formed and enriched themselves in heavy elements. I examine this scenario in some detail and in particular explore whether the searches currently being carried out to locate local examples of these MACHOs can shed any light at all on this scenario.

1. Introduction

A conservative estimate of the mass of the Galaxy out to a distance of about 2/3 of that of the Large Magellanic Cloud is $M_G = 4 \times 10^{11}$ $M_\odot$ (Fich & Tremaine 1991). With a total luminosity in the V-band of 1.4×10^{10} $L_\odot$ (Binney & Tremaine 1987) the Galactic mass to light ratio in V (M/L_V) out to 35 kpc is ~ 30. Since normal stellar populations do not generally produce M/L_V ratios higher than about 3, this is usually taken as evidence for an important component of dark matter within an extended halo surrounding the Galaxy.

A related result is that of the MACHO microlensing experiment in the direction of the Magellanic Clouds which seems to indicate that about 20% (10–50% is the 90% CI) of the dark matter in the Galaxy is tied up in objects with masses near 0.5 $M_\odot$ (0.3–0.9 is the 90% CI; Alcock 2000). This assertion is only true if the bulk of the MACHOs are located in the halo of our Galaxy. Moreover, recent studies suggest that objects with masses from 10^{-7} to 0.02 $M_\odot$ are now largely excluded (so planets and/or brown dwarfs are not important dark matter candidates) and that a 100% MACHO halo is no longer a viable model (Alcock 2000).

MACHO masses near 0.5 $M_\odot$ are suggestive of white dwarfs although neutron stars or even primordial black holes remain as unlikely but still possible candidates. A halo currently consisting of 20% old white dwarfs implies that the precursors of these 0.5 $M_\odot$ objects would have accounted for $\sim 40\%$ of Ω_{baryon} in the Universe. This estimate comes from taking 2 $M_\odot$ as the precursor masses, $M/L_V = 1500$ for Ω_{critical}, 0.04 for Ω_{baryon} and assuming that the Galactic halo is a fair sample of the types of mass seen in the Universe. It may be that Galactic halos are baryon-rich in which case the derivation of 40% would be an over-estimate. This may also be an interesting way to constrain the masses of the precursors as clearly in this analysis they could not exceed ~ 5 $M_\odot$ without violating the observed baryonic content of the Universe. Even given all the uncertainties in these estimates, it is clear in this scenario that an important fraction of all the baryons in the Universe were funneled through this star formation mode. This will surely have implications for star formation scenarios in the early Universe and perhaps also on the manner in which galaxies were assembled and how they enriched themselves in heavy elements.

The interpretation of the microlensing results is, however, not without controversy as there are indications that at least some of the lenses may reside in the Magellanic Clouds themselves (McGrath & Sahu 2000; Sahu 2002). This would have the effect of lowering

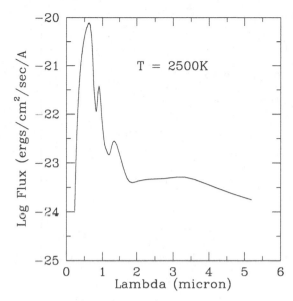

FIGURE 1. The spectrum of a 2500 K hydrogen-rich white dwarf from Hansen (1999, private communication). The depletion of flux in the near IR has been seen now in a dozen or more old white dwarfs. No known object, however, exhibits significant H_2 bands as illustrated in this model spectrum.

both the lens masses (to 0.2–0.3 $M_\odot$) and the contribution of the MACHOs to the Galactic dark matter budget. The lower lens masses would be in line with expectations from mass function considerations and the lower MACHO dark matter contribution would alleviate such problems as the appearance of high redshift galaxies (at an epoch when the white dwarf precursors were luminous), the chemical evolution of galaxies (the white dwarf precursors producing too much helium and heavier elements) and the peculiar mass function required of the precursors (a mass function truncated at both high and low masses to avoid too much chemical enrichment from high mass stars and too many halo white dwarfs currently evolving from the low mass ones; Chabrier 1999). There is also the possibility that most of the MACHOs are located in a thick (Gates et al. 1998) or flaring disk (Evans et al. 1998) of the Galaxy. This would have the effect of reducing their scale height and again their total Galactic mass contribution.

The definitive answer as to whether old white dwarfs are important baryonic dark matter contributors will come from searches which will (or will not) identify significant numbers of high velocity local examples of the very cool white dwarfs that produce the microlensing. Such searches are now underway and in the ensuing sections I will discuss early results from them.

2. New models of very cool white dwarfs

Just recently, a new era opened in the study of old white dwarfs. Hansen (1998, 1999) first calculated emergent spectra from cool ($T_{eff} < 4000$ K) white dwarfs with atmospheres that included opacity from the H_2 molecule. Similar models were also constructed by Saumon and Jacobson (1999) and earlier ones by Bergeron, Saumon & Wesemael (1995) for objects down to 4000K. All these models exhibited very strong collisionally induced opacity in the red and near IR part of the spectrum and the effects of this opacity can be seen clearly in the model spectrum shown in Figure 1 as well in

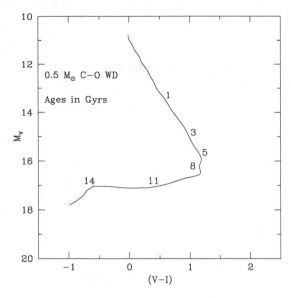

FIGURE 2. Cooling track of a 0.5 $M_\odot$ white dwarf (Hansen 1999). The age of the white dwarf in Gyrs is indicated along the plot. This age does not include the main sequence lifetime of the white dwarf precursor.

several dozen real objects (see e.g. Oppenheimer et al. 2001b). The broad bands seen in the spectrum in Figure 1 are due to H_2 but it is of interest to point out that no cool white dwarf has yet been observed that actually shows these spectral features. Perhaps it is because no objects this cool have been discovered up until now, or that the models are still somehow deficient, or that possibly the Universe is not old enough for hydrogen-rich white dwarfs to have reached such low temperatures. The effects of collisionally induced absorption are not limited, however, to extremely cool objects. Its influence can been seen in white dwarfs at least as hot as 5400 K (see, e.g. the spectrum of LHS 1126 in Bergeron et al. 1994).

The effect of the H_2 opacity is to force the radiation out in the bluer spectral regions. This has an enormous influence on their colors, the location of the white dwarf cooling track in the color-magnitude diagram, and hence their observability. Down to a temperature of about 4000 K (which corresponds to an age of about 7 Gyr for 0.5 $M_\odot$ white dwarfs) these stars become increasingly redder as they cool, the reddest color they achieve is about $V - I \simeq 1.2$ in Hansen's (1998, 1999) models. Note that other sets of cooling models give somewhat different colors and ages here (e.g. Fontaine et al. 2001). Older white dwarfs become progressively bluer in $V - I$ as they continue to cool. Ancient white dwarfs, of age 12 Gyr, and mass 0.5 $M_\odot$, have $V - I \sim -0.3$ according to these models. As has been noted several times in the past few years, *old hydrogen-rich white dwarfs are blue, not red, in $V - I$ colors.* Searches for such ancient objects are now concentrating on bluish objects in these colors.

A cooling track from Hansen's (1998, 1999) models illustrating all these features for 0.5 $M_\odot$ white dwarfs is shown in Figure 2. Note that instead of getting arbitrarily faint with age, old hydrogen-rich white dwarfs seem eventually to coast at approximately constant M_V for ages between about 10 and 14 Gyr, just the timescales which may bracket globular cluster ages. The implication is clear—very old white dwarfs may be easier to detect than was previously thought. Also potentially of interest is that the $V - I$

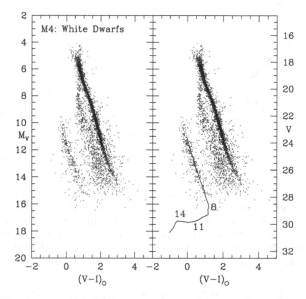

FIGURE 3. Deep HST color magnitude diagram of the globular cluster M4 with a cooling track for a 0.5 $M_\odot$ white dwarf superimposed. The oldest white dwarfs detected are only about 7 Gyr and while these should show some effects due to collisionally induced absorption, they are too warm to exhibit the extreme effect of a bluing in the $V - I$ color with decreasing temperature.

color now moves the bulk of the ancient white dwarfs away from the swarm of field K and M-dwarfs (making confusion with such objects *less* likely) but at the same time overlapping the colors of distant star-forming galaxies (making the confusion with such objects *more* likely). The solution here will be to search for old white dwarfs using both color and proper motion criteria, an approach taken by most current programs.

One problem with these models is that they are largely untested. Of serious concern is that none of the cool white dwarfs found to date (see following sections) seem to exhibit the extreme blue colors predicted by most of these current models containing H_2 in their atmospheres. This may be suggesting missing or inadequate physics. One place where these models could potentially be tested is among the cool white dwarfs in globular clusters. Such systems should contain white dwarfs almost as old as the clusters themselves (likely in the 12–14 Gyr range), so the extreme effects of collisionally induced opacity ought to be clearly seen. Figure 3 shows the current situation here. It displays the deepest color magnitude diagram yet obtained for a globular cluster (Richer et al. 1995, 1997), an HST study of M4. White dwarfs perhaps as faint as $M_V = 16$ are recorded in this study. If these objects have masses near 0.5 $M_\odot$, they will have ages in excess of 7 Gyr and $T_{\rm eff} \sim 4000$ K in Hansen's (1998, 1999) models; cool enough to exhibit some of the effects of collisionally induced opacity but not old enough to help constrain the appearance of truly ancient white dwarfs. In HST cycle 9, a group of colleagues and I were awarded time on HST to attempt to locate the termination point of the white dwarf cooling sequence in M4. With the current generation of models and a cluster age in the range of 12–14 Gyr, this should occur near a V magnitude of about 30, certainly a enormous challenge. It is not clear whether the data will actually go this deep, but even if it does not, we should be in a position to provide some stringent tests of the current generation of models.

In Figure 4 I attach a simulation of what we expect to see with these data. This is an optimistic simulation in that it does not include the effects of charge transfer inefficiency

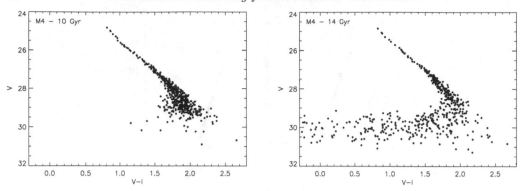

FIGURE 4. Simulation of deep HST data in the globular cluster M4. The WFPC2 exposure time calculator was used to determine the photometric uncertainty as a function of apparent magnitude for the white dwarfs in the globular cluster M4 assuming ages of 10 and 14 Gyr. A suite of Monte Carlo realizations, incorporating a proper treatment of the photometric uncertainties involved and the higher background expected from scattered light of the bright giants, were then undertaken. One such realization is shown for a cluster with each of these ages.

now known to be plaguing the HST WFPC2 CCDs and also assumes that all the data are taken under low sky brightness conditions. Nevertheless, even if the faintest magnitudes are not reached, the effects of collisionally induced opacity should be readily apparent and quantitatively testable with these data.

3. Current Searches for Very Cool White Dwarfs

A number of searches for an ancient population of halo white dwarfs are currently underway or have been completed. Most of these involve color and proper motion selection with some follow-up spectroscopy. The proper motions have generally been derived from digitized photographic plates with quite long time baselines (typically 20–50 years). I list below in Table 1 a compilation of these surveys together with some additional information. This table must be considered a rough guide only to the current situation. However, it is indicative of the present state of our knowledge.

In Table 1 the first two columns give the name and the area searched for each program. The third column is the limiting V magnitude of each survey. This is a fairly crude estimate as some (e.g. Oppenheimer et al. 2001a) did not survey in the V-band and it was necessary to transform their particular magnitude to V so that a homogeneous set of statistics for each cluster could be derived. Column 4 is an estimate of the distance probed in each survey for old white dwarfs. It is the distance out to which ancient white dwarfs could have been detected in the individual surveys. This limiting distance for detection of old white dwarfs was set to $10(10^{([V_{lim}-17]/5)})$ pc; 17 was taken as the M_V of a typical old white dwarf (Hansen 1998, 1999; Saumon and Jacobsen 1999; Richer et al. 2000). Column 5 lists the number of thick disk and spheroid white dwarfs that are expected out to the limiting distance of the survey. This number was derived following the prescription of Reid et al. (2001) and is dominated by the thick disk contribution. Thin disk white dwarfs are not considered in this discussion as they would be eliminated by their low velocities. We are only concerned here with high velocity objects which could contaminate a true dark halo sample. Column 6 contains the number of observable white dwarfs expected from the dark halo (Richer et al. 2000) assuming that the halo consists of 20% by mass of MACHOs which are old 0.5 $M_\odot$ white dwarfs, half of which have hydrogen-rich outer atmospheres. The helium-rich ancient white dwarfs would have

Survey	Area (Deg²)	V_{lim}	D_{max} (pc)	No. Disk + Sph.	No. Dark Halo	No. Found
Opp†	4165	19.8	36	3.9	3.7	4
Monet‡	1378	20	40	2.5	2.4	>1
Ibata¶	790	20	40	1.4	1.4	2
SDSS‖	400	20.5	50	1.4	3.5	2
Jong††	2.5	24	250	1.1	3.6	3
CFHT‡‡	15	25	400	27	88	?
HDF¶¶	1.4×10^{-3}	28	1600	0.2	0.5	0?

† Oppenheimer et al. 2001a, Science Online, March 22
‡ Monet et al. 2000, AJ 120, 1541
¶ Ibata et al. 2000, ApJ 532, L41
‖ Harris et al. 2001, ApJ 549, L109
†† de Jong et al. 2000 astro-ph/0009058
‡‡ Stetson et al. in progress
¶¶ Ibata et al. 1999, ApJ 524, L95

TABLE 1. Predicted and observed numbers of faint white dwarfs

cooled to very low luminosities by a Hubble Time and would be basically unobservable. The seventh column lists the number of old white dwarfs actually found **out to the limiting distance only** in each survey.

It is clear from examination of Table 1 that in all existing surveys thus far, there is no obvious excess in the discovered number of old white dwarfs over that expected from the thick disk and spheroid. Note, however, that the CFHT program, currently acquiring its second epoch of data, should be extremely important in this context. It is quite different from most of the other surveys as it is much deeper ($V_{lim} \sim 25$). One caveat must be kept in mind when perusing Table 1, however. In comparing, for example, columns 5 through 7 the assumption of 100% recovery of all moving objects is built in. If, for example, a survey was only 50% complete in recovering moving objects, then the number in column 7 should be increased by a factor of 2. For some surveys, such as the SDSS, where proper motions are not part of the selection criteria, the numbers in Table 1 are strictly correct. However, until estimates for proper motion limits are accurately known, there remains no clear evidence of a white dwarf dark matter halo signature in any of the current surveys. Of course, at the moment, it is also true that the surveys do not rule out such a dark matter halo either.

4. Temperature and age distribution of dark halo white dwarf candidates

We can examine the totality of objects in the surveys of Table 1 from a different perspective. If they are, in fact, local examples of the MACHOs, and thus important Galactic dark matter contributors, they should be quite old, likely as old as the globular clusters. This would place them in the 12–14 Gyr range if we take the current spread quoted for globular cluster ages. From the location of the white dwarfs in color-magnitude and color-color diagrams, temperatures can be inferred with the use of theoretical models. From these temperatures, ages can then be assigned using models of cooling white dwarfs after choosing an appropriate mass.

For the purposes of this exercise, I used the models of Hansen (1998, 1999) and those discussed in Oppenheimer et al. (2001a) with white dwarf masses of 0.5 $M_\odot$, to estimate the temperature and cooling time of each object found in the various surveys. Note that

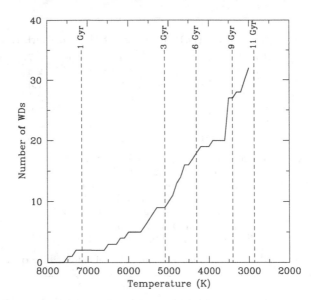

FIGURE 5. Cumulative distribution of the temperatures of all the white dwarfs with published colors from the surveys listed in Table 1. Here I have selected not just white dwarfs out to D_{max} for each survey, but the complete sample. The time for a 0.5 $M_\odot$ white dwarf to cool to a given temperature is indicated. The age distribution seems to come from a sample of objects that is significantly younger than the halo.

we are not limiting ourselves to objects within D_{max} here. The cumulative distribution of temperatures is shown in Figure 5 and the time for a 0.5 $M_\odot$ white dwarf to cool to a given temperature is also indicated by the vertical lines. The immediate conclusion from the plot is that the sample in Table 1 is *not* that of a halo sample, the objects are simply too young. It seems more in line with that of a thick disk component which dominates the predicted numbers in Table 1. This result is, however, strongly model dependent. A different set of cooling models or mass would provide quite a different result. For example, if I had chosen to assign a mass of 0.6 $M_\odot$ to all the stars, the coolest object would have a cooling age of 11.7 Gyr, almost 1 Gyr older than a 0.5 $M_\odot$ white dwarf. Further, the main sequence lifetimes of the white dwarf precursors are not included in the cooling ages. If the masses of the precursors are in the range of 2 $M_\odot$ this will add about 1 Gyr to these numbers. Nevertheless, the general conclusion to be drawn from both Table 1 and Figure 5 is that within the context of the present generation of white dwarf cooling models, truly ancient white dwarfs do not currently appear to have been discovered.

5. Moving objects in the Hubble Deep Field

The last row in Table 1 deserves some special comments. Based on two epochs of exposures taken in the HDF separated by 2 years, Ibata et al. (1999) claimed to detect two extremely faint objects with significant proper motion. One of these appeared to have colors consistent with that expected from a very cool white dwarf. With lower statistical significance, three other objects also appeared to be moving, one of which possessed colors expected from an old white dwarf. The inferred space velocities for these objects placed them squarely in the category of halo objects. This result was potentially very important as the detection of even a few such objects in the HDF would be enough to account for the entire MACHO contribution to the Galactic dark matter. The second epoch data were, however, taken in the non-optimal F814W filter for another program (supernova search), and were quite noisy.

Based on this seemingly important result, a third epoch was secured in F606W, fully five years after the original HDF data. These images were of significantly higher quality. One of the objects suggested by Ibata et al. (1999) as possibly moving was number 4-141 in the Williams et al. (1996) compilation. Using a maximum likelihood technique, Ibata et al. (1999) measured a motion of 35 mas/yr with an error of 8 mas/yr. With the current superior data set we measure a proper motion of 10 ± 4 mas/yr, a proper motion which is not statistically significant. None of the other 4 objects was found to possess significant motion over the 5-year baseline.

A full discussion of our reanalysis of moving objects in the HDF will appear in Richer et al. 2002.

6. A final comment

My personal view of the current landscape regarding the viability of old white dwarfs as a component of the Galactic dark matter is that while there are some interesting hints, there is no compelling evidence in the current statistics either for against the scenario. Perhaps a listing of the critical issues will be instructive.

(a) *Where are the lenses in the direction of the LMC and SMC located?* This is a fundamental issue—perhaps the most important one. There are suggestions that some, perhaps even most, of the lenses are in the Clouds themselves. This issue may be settled in a number of ways, some of which are possible with current technology. One road to a solution is to observe along a different line of sight—a microlensing campaign in the direction of M31 or M33 would be extremely useful here. There are programs currently underway looking at M31 (e.g. Uglesich et al. 1999) but no comprehensive results are available at this time. Some future satellite missions may be capable of detecting parallaxes (and masses) for the lenses (Gould and Salim 1999) and this would be critical data for establishing their location. If the lenses are in the Clouds and are low mass main sequence stars, then it may eventually be possible to observe them directly.

(b) *Is there a population of cool white dwarfs with halo kinematics that are present in numbers which exceed those expected from the thick disk and spheroid?* As I have discussed, based solely on number count arguments, the current searches do not as yet provide a conclusive answer either in the affirmative or negative. Deep (V $\sim$ 25) proper motion and color selected surveys covering 10 degrees or more of sky are required to be definitive. There are several such searches underway and the results are eagerly anticipated.

(c) *Do we have the last word on the models of old white dwarfs?* There are two aspects of the present theory of cooling white dwarfs which I find disturbing. First, there are enormous problems in fitting the spectra of these cool white dwarfs with the current generation of models. The situation is so dire that Bergeron (2001), one of the premier practitioners of this art, was required to insert a fictitious opacity source into his models to help in the fits. He then went on to state: "It now remains to be seen whether this new opacity source, even when included with the approximate treatment introduced in this paper, can help resolve the mystery surrounding several ultracool white dwarfs whose energy distributions have yet failed to be successfully explained in terms of hydrogen or mixed hydrogen/helium compositions." The second concern is simply that no truly ultracool white dwarfs have as yet been discovered, or, perhaps better put, no white dwarfs are known whose colors and/or spectra suggest that the object's age is in the range of 12–14 Gyrs. This may be the same problem as the first one mentioned here, that is that the models are still somehow inadequate, or it may be suggesting that such objects are extremely rare or just do not exist. It is within this context that extremely deep observations of globular cluster white dwarfs will be important.

The author would like to thank the Space Telescope Science Institute for organizing a meeting that was all that a scientific should be; exciting new results, participants with deep and wide interests, an excellent venue. In particular, deepest gratitude is due to Mario Livio who organized the meeting and provided one of the best conference summaries that this author has ever experienced.

The research of HBR is supported by grants from the Natural Sciences and Engineering Research Council of Canada.

REFERENCES

ALCOCK, C. 2000 *Science* **287**, 74.

BERGERON, P. 2001 *ApJ* **558**, 369; astro-ph/0105333.

BERGERON, P., RUIZ, MARIA-TERESA, LEGGETT, S. K., SAUMON, D., & WESEMAEL, F. 1994 *ApJ* **423**, 456.

BERGERON, P. SAUMON, D., & WESEMAEL, F. 1995 *ApJ* **443**, 764.

BINNEY, J. & TREMAINE, S. 1987 *Galactics Dynamics*, Princeton Series in Astrophysics, Princeton University Press.

CHABRIER, G. 1999 *ApJ* **513**, L103.

DE JONG, J., KUIJKEN, K., & NEESER, M. 2000; astro-ph/0009058.

EVANS, N. W., GYUK, G., TURNER, M. S., & BINNEY, J. 1998 *Apj* **501**, L45.

FICH, M. & TREMAINE, S. 1991 *ARAA* **29**, 409.

FONTAINE, G., BRASSARD, P., & BERGERON, P. 2001 *PASP* **113**, 409.

GATES, E. I., GYUK, G., HOLDER, G.P., & TURNER, M.S. 1998 *ApJ* **500**, L145.

GOULD, A. & SALIM, S. 1999 *ApJ* **524**, 794.

HANSEN, B. M. S. 1998 *Nature* **394**, 860.

HANSEN, B. M. S. 1999 *ApJ* **520**, 680.

HARRIS, H. C., HANSEN, B. M. S., LIEBERT, J., VANDEN BERK, D. E., ANDERSON, S. F., KNAPP, G. R., FAN, X., MARGON, B., MUNN, J. A., NICHOL, R. C., PIER, J. R., SCHNEIDER, D. P., SMITH, J. A., WINGET, D. E., YORK, D. G., ANDERSON, J. E., JR., BRINKMANN, J., BURLES, S., CHEN, B., CONNOLLY, A. J., CSABAI, I., FRIEMAN, J. A., GUNN, J. E., HENNESSY, G. S., HINDSLEY, R. B., IVEZIC, Z., KENT, S., LAMB, D. Q., LUPTON, R. H., NEWBERG, H. J., SCHLEGEL, D. J., SMEE, S., STRAUSS, M. A., THAKAR, A. R., UOMOTO, A., & YANNY, B. 2001 *ApJ* **549**, L109.

IBATA, R., IRWIN, M., BIENAYME, O., SCHOLZ, R., & GUIBERT, J. 2000 *ApJ* **532**, L41.

IBATA, R., RICHER, H. B., GILLILAND, R. L., & SCOTT, D. 1999 *ApJ* **524**, L95.

McGRATH, E. J. & SAHU, K. C. 2000 *AAS Meeting 197*, #105.09.

MONET, D. G., FISHER, M. D., LIEBERT, J., CANZIAN, B., HARRIS, H. C., & REID, I. N. 2000 *AJ* **120**, 1541.

OPPENHEIMER, B. R., HAMBLY, N. C., DIGBY, A. P., & SAUMON, D. 2001a *Science* **292**, 698.

OPPENHEIMER, B. R., SAUMON, D., HODGKIN, S. T., JAMESON, R. F., HAMBLY, N. C., CHABRIER, G., FILIPPENKO, A. V., COIL, A. L., & BROWN, M. E. 2001b *ApJ* **550**, 448.

REID, I. N., SAHU, K. C., & HAWLEY, S. L. 2001 *ApJ*, **559**, 942; astro-ph/0104110.

RICHER, H. B., ET AL. 2002, in preparation.

RICHER, H. B., FAHLMAN, G. G., IBATA, R. A., PRYOR, C., BELL, R. A., BOLTE, M., BOND, H. E., HARRIS, W. E., HESSER, J. E., HOLLAND, S., IVANANS, N., MANDUSHEV, G., STETSON, P. B., & WOOD, M. A. 1997 *ApJ* **484**, 741.

RICHER, H. B., FAHLMAN, G. G., IBATA, R. A., STETSON, P. B., BELL, R. A., BOLTE, M., BOND, H. E., HARRIS, W. E., HESSER, J. E., MANDUSHEV, G., PRYOR, C. & VANDENBERG, D. A. 1995 *ApJ* **451**, L17.

RICHER, H. B., HANSEN, B., LIMONGI, M., CHIEFFI, A., STRANIERO, O., & FAHLMAN, G. G. 2000 *ApJ* **529**, 318.

SAHU, K. C. 2002; see paper in this volume.

SAUMON, D. & JACOBSON, S. 1999 *ApJ* **511**, L107.

UGLESICH, R., CROTTS, A. P. S., TOMANEY, A. B., & GYUK, G. 1999 *AAS Meeting 194*, #90.04

WILLIAMS, R. E., BLACKER, B., DICKINSON, M., DIXON, W., FERGUSON, H. C., FRUCHTER, A. S., GIAVALISCO, M., GILLILAND, R. L., HEYER, I., KATSANIS, R., LEVAY, Z., LUCAS, R. A., McELROY, D. B., PETRO, L., POSTMAN, M., ADORF, H-M., & HOOK, R. 1996 *AJ* **112**, 1335.

Hot gas in clusters of galaxies and Ω_M

By MEGAN E. DONAHUE

Space Telescope Science Institute, 3700 San Martin Drive, Baltimore, MD 21218 USA

X-ray clusters provide excellent constraints on cosmological parameters such as Ω_M. I will describe measurements of cluster masses and of cluster evolution. The cluster baryon fraction and the evolution of the cluster temperature function strongly constrain the mean density of matter in the universe (Ω_M). The constraints are consistent with $\Omega_M = 0.2$–0.5, with best fit values of $\Omega_M = 0.3$–0.4. The systematic uncertainties are of the same size as the statistical uncertainties, even with the small number of clusters in our current temperature surveys ($\Delta\Omega_M \sim 0.1$.) Thus, reduction of the uncertainties in these methods requires not only an increased number of hot massive clusters in a given sample but much better quantification of the systematics, a goal which demands not only more clusters but clusters with a range of properties and redshifts. The current constraints are not particularly sensitive to the particular form or value of the acceleration parameter Λ and therefore these constraints provide an limit on cosmological parameters complementary to the limits imposed by the cosmic microwave background studies and by the Type Ia supernovae at cosmological distances.

1. Introduction

I seek to make the following three points in this review:

(a) Clusters of galaxies are excellent targets for cosmological studies.

(b) Existing studies have already placed very strong constraints on the mean density of matter in the universe.

(c) These constraints are nearly orthogonal to constraints from the cosmic microwave background and type Ia supernovae.

(d) Future studies will place extremely powerful limits on cosmological parameters.

Clusters of galaxies are the largest relaxed gravitating systems in the universe. The various means of measuring the masses of these large systems are all statistically consistent (with some outliers); I will show that the evidence is fairly good that the most massive of the cluster systems are nearly in gravitational equilibrium punctuated by occasional interactions and mergers with other systems. A massive cluster is permeated by a hot ($kT \sim 2$–12 keV), dense ($n_{core} \sim 10^{-2}$–10^{-3} cm^{-3}), X-ray emitting atmosphere, the intracluster medium (ICM) confined by the cluster's gravitational potential. The properties of the X-ray gas have recently been confirmed by spectacular Sunyaev-Zel'dovich (SZ) imaging (e.g. Grego et al. 2001). SZ imaging reveals the imprint of the cluster against the cosmic microwave background left when the electrons in the hot gas scatter the cosmic microwave background photons (Figure 1). Measurements of the S-Z effect confirm the presence and amount of the hot gas. The X-ray emitting gas outweighs the mass associated with galaxies (including the galaxy's dark matter) in the rich clusters by at least a factor of 3–5 (e.g. David et al. 1990; Edge et al. 1991; Evrard 1997; Roussel, Sadat & Blanchard 2000).

The global properties of the ICM in any given cluster are governed primarily by the dark matter content of that cluster. Indeed, the diffuse intergalactic gas left over from galaxy formation (which is most of the baryons), may trace the gravitational potential of the filaments of dark matter (See the hydrodynamical simulation from Bryan & Voit (2001) reproduced in Figure 2). Sufficiently sensitive X-ray images of the extended thermal component of the X-ray background may reveal a web of structure. (In actual X-ray imaging observations, extragalactic point sources, AGN at the high flux levels and in-

MS 0451−03: S−Z Effect Contours, Chandra ACIS Color Scale

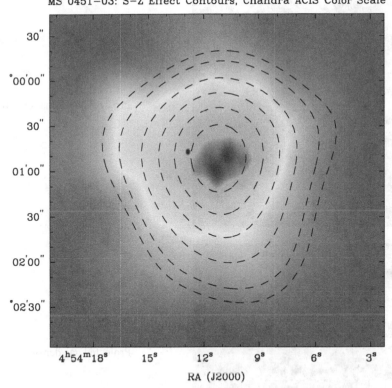

FIGURE 1. Chandra imaging (Donahue et al., in prep.) and Sunyaev-Zel'dovich imaging from Owens Valley Radio Observatory (Reese et al. 2000) of the $z = 0.54$ cluster MS0451−0302. Figure courtesy Jessica Gaskin and Sandeep Patel (MSFC).

creasingly normal galaxies at the lower flux levels, also trace the large scale structure, but in a less predictable fashion because of the uncertainties in the processes included in AGN and galaxy formation.)

2. X-ray properties of clusters

The properties of clusters, their X-ray properties in particular, make them excellent cosmological probes.

As X-ray sources with luminosities of 10^{43}–10^{46} erg s^{-1}, clusters outshine nearly every other X-ray source in the universe except for the brightest quasars, a surprising result to X-ray astronomers in the 1960s and early 1970s (e.g. Boldt et al. 1966; Gursky et al. 1971). Images of X-ray clusters span arcminutes, co-spatially with the cluster galaxies. Clusters are centrally concentrated, with core radii of $\sim 0.12h^{-1}$ Mpc where the Hubble constant H_0 is parameterized as $h = H_0/100$ km s^{-1} Mpc^{-1}. X-ray emission from hot gas has been detected from nearby clusters at projected radii as large as 5 Mpc. The diffuse gas is optically thin at most wavelengths, and therefore derivations of gas masses are not complicated by the effects of radiative transfer. However, while the detection of clusters of galaxies at cosmological distances is not extremely hard, the detection of the extended emission from more distant clusters is limited by the surface brightness dilution factor of $(1 + z)^{-4}$, making the detection of a distant cluster out to its virial radius a very difficult proposition.

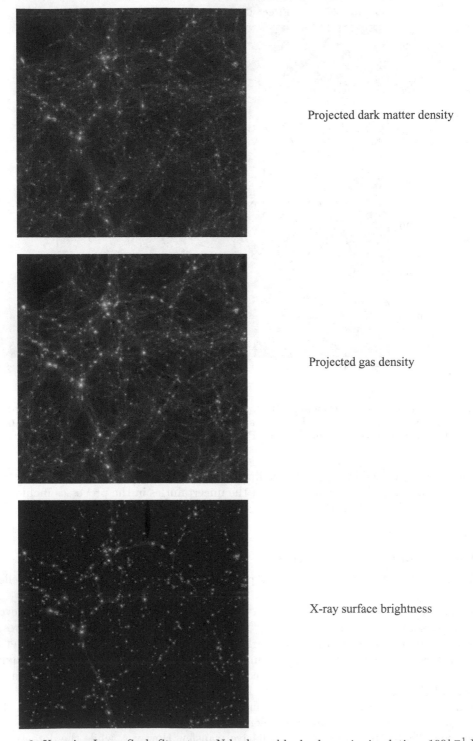

Projected dark matter density

Projected gas density

X-ray surface brightness

FIGURE 2. X-raying Large Scale Structure. N-body and hydrodynamic simulations $100h^{-1}$ Mpc on a side show similar structures in the dark matter and the gas, and that the X-ray emission traces the highest density gas ($\propto n^2$) (Bryan & Voit 2001).

The intracluster gas of an X-ray cluster radiates via thermal bremsstrahlung (free-free emission) and collisionally excited emission lines from ions which still retain some electrons. The hottest clusters emit nearly 100% thermal bremsstrahlung continuum with Fe K line emission from nearly fully ionized iron. The temperature from such a gas can be determined from the spectral shape of the continuum with a characteristic high energy cut-off at $\sim kT$. Temperatures of clusters' ICM are found to range from 1–15 keV (or $10^7 - 2 \times 10^8$ K).

Cluster ICM temperatures are very strongly correlated with cluster X-ray luminosities, which typically range between 10^{43}–10^{46} erg s^{-1} bolometric. The $L_x - T_x$ correlation arises naturally from a physical relationship between the temperature and the gravitational potential of the cluster, the fraction of baryons in the gas phase, and the minimum entropy of the gas. (Explaining the relationship in detail is the goal of current cluster formation simulations and goes beyond the scope of this article.)

The combination of the density $n_e(r)$ and temperature $T(r)$ profiles from clusters as obtained by observations with sufficient signal, spectral resolution, and spatial resolution, leads to a pressure profile. A pressure profile $P(r)$ leads to $M(r)$ directly if one assumes that the gas is in nearly hydrostatic equilibrium with the gravitational potential of the cluster.

Outside the core, clusters often look regular or relatively relaxed. Recent high spatial-resolution Chandra observations of clusters of galaxies reveal that the cores of these clusters are affected by moderate sized radio sources and other processes. However, these high resolution images as well as lower resolution and wider-field images also demonstrate that while the core may be irregular, the more distant isophotes are not (McNamara et al. 2000)

Within a few core radii, the distribution of gas within a cluster of galaxies is well-modelled by a "β-function." The surface brightness of the X-ray cluster can be fit to a radial distribution profile $\propto [1 + (r/r_c)^2]^{-3\beta+1/2}$. Recent XMM observations suggest that the gas is very nearly isothermal inside the region defined by the X-ray emission. For example, Arnaud et al. (2001) report that Abell 1795 is isothermal out to > 0.4 virial radii. A deprojection of the β-model assuming nearly isothermal gas then results in a physical gas density distribution $\propto [1 + (r/r_c)^2]^{-3\beta/2}$.

Many astronomers outside the X-ray community are surprised to learn that baryons in the form of hot gas outweigh the matter associated with the galaxies by a large amount. The X-ray gas in clusters outweighs the stars in the same clusters by a factor $M_g/M_* \sim 5$–$10h^{-3/2}$ (David et al. 1990; Edge et al. 1992; Arnaud et al. 1992; Roussell, Sadat & Blanchard 2000). Indeed, the "M_*" used in these ratios are based on assumptions regarding the galaxies M/L ratios, with a conservatively high M/L_B ratio typically $\sim 6h$ in solar units (van der Marel 1991). Such a mass to light ratio is chosen to be typical of spheroidal star systems, and includes not only the stellar mass but also all of the dark baryons associated with the galaxy systems, like stellar remnants, planets, cold gas, MACHOS (Fukugita, Hogan & Peebles 1998). *The mass ratio of gas to stars is intriguing because it implies not only that most of the baryons in the universe are NOT in stars but in the gas phase, and also that most of the baryons in the universe may never have been cycled through a star.*

We can show that the gravitational masses of clusters estimated by independent means are self-consistent. Clusters of galaxies have the unique advantage of having masses measurable by at least three independent means. Each method has its limitations, but the methods complement each other. X-ray estimates rely on the gas being in equilibrium; weak lensing estimates measure the mass, but in a cylinder along the line of sight and thus depend on geometry; velocity dispersions of galaxies can be contaminated by non-member galaxies and large scale structure effects.

Weak lensing masses for clusters are consistent with the masses estimated from X-ray observations outside the core. Wu et al. (1998) studied 38 clusters comparing the weak lensing masses with the optical velocity dispersions and found good consistency. Soucail et al. (2000) found that the cluster CL0024+17, which has an X-ray-determined mass lower than its optical and weak lensing masses, is a system with a filament along the line of sight. Many other studies find consistency between the X-ray determined mass, the optical velocity dispersion and the weak lensing masses in individual high redshift systems (e.g. Donahue et al. 1998; Tran et al. 1999; Hjorth et al. 1998; Bohringer et al. 1998) and in larger low-redshift samples (Allen 1998). Such consistency implies that even if the clusters occasionally depart from their equilibrium states, the departures are either not large and/or the departure times are very short on cosmological timescales. Statistically, the mass estimates from cluster temperatures, weak lensing observations, and galaxy velocity dispersions are consistent with each other, and therefore cluster temperatures alone are likely to be a reliable predictor of a cluster's gravitating mass.

Thus for a number of reasons, clusters are excellent cosmological probes.

(a) As I have described in this section, clusters appear to be accessible, relatively simple physical systems (in contrast to the vast uncertainties associated with the formation of galaxies and AGN). Whether this accessibility is a fortuitous feature of our universe or a required feature is beyond the scope of this article.

(b) Clusters are very massive, and form from such a large volume they are thought to provide a "fair sample" of the universe.

(c) Clusters are very massive, and thus their formation and evolution are sensitive to initial conditions and the mean density.

In the remainder of this article, I will explore the implications of the items b and c in this list.

3. Method 1: Ω_M from cluster gas fractions

Since clusters of galaxies form from a volume of space which is very large (~ 10 Mpc), the most massive clusters of galaxies are thought to be fair samples of the universe. Thus the baryon fraction or the M/L ratio measured in clusters of galaxies are expected not only to be the same from cluster to cluster, but also to be representative of those values averaged over the entire universe.

From this fair sample hypothesis, we can use the baryon fraction of clusters $f_b = \Omega_b/\Omega_M$ and the value of Ω_b from primordial nucleosynthesis models and constraints on the primordial abundance of deuterium from quasar absorption line studies to infer Ω_M. To my knowledge, the basic idea to infer Ω from cluster gas fractions and constraints from primordial nucleosynthesis was first published by Briel, Henry & Bohringer (1992) in their paper on ROSAT observations of the Coma Cluster; the idea was further developed by the more well-known paper by White, Navarro, Evrard & Frenk (1993) who tested hydrodynamic simulations of clusters for a correction to the pure fair sample hypothesis and adjusted the gas fraction to the proper baryon fraction by estimating the minor contribution of mass associated with galaxies (stars and dark matter). Both of those studies as well as later analyses (e.g. Evrard 1997) inferred that Ω_M was highly inconsistent with 1, a conclusion which was known as the "baryon catastrophe" in the 1990s.

Estimates of the hot gas fraction from X-ray and S-Z observations are consistent. Mohr et al. (1999) found hot gas fractions for a sample of 45 clusters using two different methods (with corrections for clumping and depletion) to be $f_g = 0.075 \pm 0.010h^{-3/2}$, similar to Evrard's (1997) result ($f_g = 0.06 \pm 0.003h^{-3/2}$). Grego et al. (2001) found a very similar result using beautiful Sunyaev-Zel'dovich observations for the gas contribution and X-ray observations for the gravitating mass contribution to the gas fraction $f_g = 0.08 \pm 0.01h^{-1}$.

These gas studies are consistent with $\Omega_M \lesssim 0.25h^{-1/2}$ (or $0.3h^{-1}$), if $\Omega_B h^2 = 0.019 \pm 0.001$ from baryonic nucleosynthesis and a primordial deuterium measurement (Burles & Tytler 1998) or if $\Omega_b h^2 = 0.022 \pm 0.005$ as estimated from the BOOMERANG experiment (de Bernardis et al. 2002).

4. Method 2: Cluster temperature function evolution and Ω_M

One can use the cluster temperature function and its evolution to constrain Ω_M. The *rate* of structure formation depends sensitively on Ω_M. By comparing the number of hot, massive clusters at high redshift with the number of clusters at $z \sim 0$, we can constrain Ω_M while being relatively insensitive to Ω_Λ or Λ (depending on your preferred notation).

One may also use the evolution of the cluster luminosity function to constrain Ω_M, and I refer the reader to Rosati's contribution in this same volume.

Cluster evolution simulations show a dramatic evolution in the number density of hot clusters from $z \sim 1$ to $z \sim 0$, with very few clusters at high redshift relative to low redshift. In contrast, the low density models show minimal amounts of evolution in the number density of such clusters. The physical explanation for this result is that in a critical density universe, Ω_M starts at 1 and stays at 1 for all time. In such a universe, the construction of large scale structure occurs continuously. In a low-density universe, the mean density of the universe as parameterized by Ω_M started out near 1, but then at the time corresponding very approximately to the redshift $z \sim \Omega^{-1} - 1$, the mean mass density departed from one and structure formation slowed considerably. (A more precise expression for the time at which structure formation begins to slow in an open universe is $t \sim \pi H_0^{-1}\Omega_M/(1-\Omega_M)^{3/2}$.) Therefore in a low density universe, large scale structure at $z \sim 1$ looks rather similar to the LSS at $z \sim 0$—structure formation hasn't precisely stopped, but slowed considerably.

Evaluation of the cluster temperature function reveals that its evolution is very sensitive to the density parameter, particularly for the hottest clusters. Lilje (1992) and Oukbir & Blanchard (1992) had calculated that the X-ray temperature function and its evolution could place constraints on Ω_M. But there was no temperature sample with sufficient numbers of clusters at high redshifts to perform such an analysis until temperature measurements of the Extended Medium Sensitivity Survey (EMSS; Gioia et al. 1990ab) were made available (Henry 2000; Donahue et al. 1999).

The EMSS was a survey of X-ray sources serendipitously discovered in EINSTEIN/IPC observations. Such a study was the first of its kind, made possible by the imaging capability of the EINSTEIN/IPC, the first ever X-ray imaging system. The X-ray images revealed hundreds of new X-ray sources. An extensive identification campaign led by Stocke et al. (1991) turned up the largest sample of X-ray clusters, over 100 clusters at redshifts between 0.14–0.83. In addition, the detection volume for these clusters was quantifiable based on the selection algorithm and the surface brightness limit of the survey. The high number of high redshift clusters and the quantifiable volume of the survey made it possible for the first time to make reliable cosmological comparisons. The now famous cluster MS1054−0321 was discovered in a IPC field pointed at a very faint quasar. Ground-based follow up revealed a cluster at $z = 0.828$. ASCA observations revealed an X-ray temperature of 12 ± 2 keV (Donahue et al. 1998), one of the hottest clusters known.

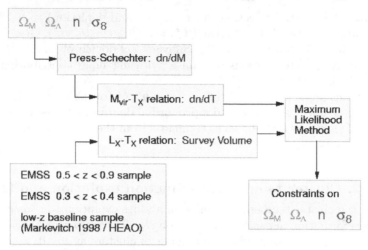

FIGURE 3. Maximum Likelihood Analysis

The mere existence of such a cluster at $z = 0.8$ in the EMSS area severely challenged the assumption of $\Omega_M = 1$. Even with the revised temperature of 10.5 ± 2 keV from Chandra observations (Jeltema et al. 2001), this single cluster presents a dilemma for $\Omega_M = 1$ models of large scale structure.

But MS1054−0321 was not alone. There were 4 other clusters in the EMSS with $z > 0.5$ with temperatures determined from ASCA X-ray observations (Donahue et al. 1999). There were 9 others between $z = 0.3$–0.4. Henry (2000) published a complete listing of the EMSS clusters with ASCA temperatures between $z = 0.3$–0.6. Such a catalog plus the EMSS survey volume can be used to place constraints on Ω_M (Donahue & Voit 1999; Henry 2000; Eke, Cole, Frenk & Henry 1998).

The details of constraining cosmological parameters, which are summarized in Figure 3 can be found in the cited papers. We start with cosmological parameters σ_8 (which normalizes the initial density perturbation spectrum at the scale of $8h^{-1}$ Mpc) and a perturbation spectrum shape. In practice, Donahue & Voit (1999) use ν_c, a number representing the number of sigma corresponding to a fluctuation that grows into a cluster of 4 keV and n, the slope of the spectrum over the range of scales to which we are sensitive, over which it is approximately a power law. Over a range of Ω_M and Λ, one can derive the expectations of dn/dM, the number density as a function of mass for any given redshift range from Press Schechter (1974) with modifications from Lacey & Cole (1993). We then apply the relation between virial mass and temperature (Eke et al. 1998) to derive dn/dT and predict the number density of clusters of galaxies as a function of redshift and temperature.

On the data side, one can take the distribution of temperature and redshifts, apply the $L_x - T_x$ relation to derive a survey volume and then use a maximum likelihood method to use the unbinned distribution to place constraints on Ω_M, Λ, n, and ν_c (or equivalently σ_8). In Donahue & Voit (1999) we used two low-redshift baseline samples (Markevitch 1998 and the HEAO sample originally described in Henry & Arnaud 1992 and updated in Henry 2000), and two redshift samples from the EMSS, including the highest redshift clusters in the EMSS. Using the maximum likelihood methodology, we could examine the effects of various assumptions, including the dispersion in the relations between M_{virial} and T_x, between L_x and T_x, the evolution in such relations (including a late formation model). Another important effect was the assumption of the Press Schechter (1974)

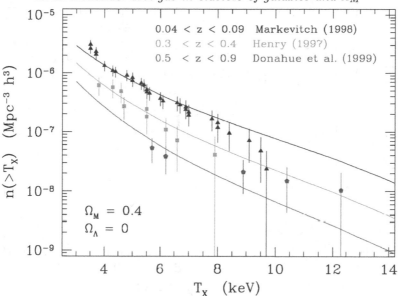

FIGURE 4. Cumulative cluster temperature function evolution for low-redshift (Markevitch 1998) and EMSS clusters through three epochs, $z = 0.04$–0.09 (triangles), $z = 0.3$–0.4 (boxes), and $z = 0.5$–0.85 (circles), plotted as data points with X-ray temperatures from ASCA and ROSAT. The error bars represent the Poisson uncertainty in each box. The width of the bin is the horizontal separation between the data points. Note that only moderate evolution is seen in the temperature function, and it is adequately fit by a flat cosmological model with $\Omega_M = 0.4$; a similar fit (represented by the smooth lines) is achieved with an open cosmological model.

formalism. Donahue & Voit (1999) experimented with a modified Press-Schechter relation that better predicted one numerical simulation. Such effects are all important to take into account, but the results were relatively robust: the conclusion that $\Omega_M < 1$ was solid at the $> 3\sigma$ level for all assumptions. The best-fit $\Omega_M = 0.4$ for open models and $\Omega_M = 0.3$ for flat models (Figure 4). Fits with $\Omega_M = 1$ were excluded with better than 3σ confidence (Figure 5).

Statistically the formal (1σ) uncertainty on Ω_M in this work was ± 0.1, with an additional systematic uncertainty of ± 0.1. The similarity between the statistical and systematic uncertainties with this method means that the addition of a few more clusters (or even a factor of three more clusters) will not improve the power of this test.

Evrard (2001) and the Virgo Consortium came to a similar conclusion in their analysis of predictions not from Press Schechter formalism, but from numerical simulations of a Hubble volume. One needs such a large simulation volume to truly test these models because of the extreme scarcity of hot clusters in an $\Omega_M = 1$ universe—the number of $T_x > 8$ keV clusters with $z > 0.5$ in an $\Omega_M = 1$ universe is ~ 100 (that is, an all-sky survey to sufficient bolometric flux limit of $\sim 5 \times 10^{-13}$ erg s^{-1} cm^{-2} would find *all* of the 8 keV clusters in the Universe (Figure 6)—fewer than 100 of them if $\Omega_M = 1$.

The constraints that both of these methods place on Ω_M are relatively insensitive to Λ, at least for clusters with $z < 1$. Therefore such constraints have a different correlation on a parameter space study of the $\Omega_M - \Lambda$ plane than does the cosmic microwave background fluctuations which constrain $\sim \Omega_M + \Omega_\Lambda$ and the Type Ia supernovae which constrain $\sim \Omega_\Lambda - \Omega_M$ (Figure 7).

One interesting implication is that there is a quantifiable flux limit above which all clusters of a given temperature will be found—meaning that deep, narrow field surveys are

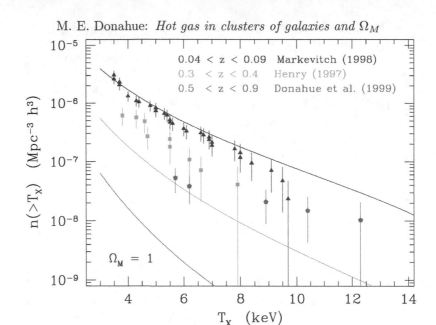

FIGURE 5. The same data as for Figure 4, but the fit is the best fit if Ω_M is constrained to be unity.

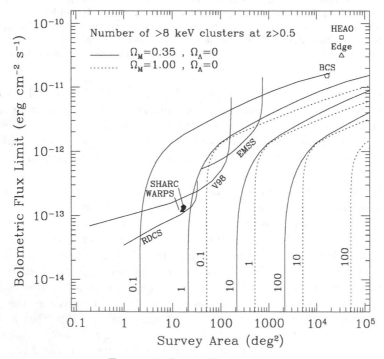

FIGURE 6. Survey Expectations

not optimal for finding significant numbers of the most massive clusters. A tuned survey would survey as large area as possible to a bolometric flux limit of $\sim 10^{-12}$ erg s^{-1} cm^{-2} (modulo the minimum temperature of the clusters one is interested in detecting—this flux is relevant for 8 keV clusters.)

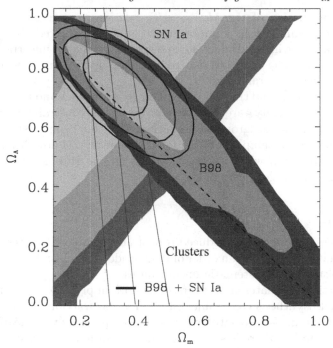

FIGURE 7. The contour plot of constraints on Ω_M and Ω_Λ from maximum likelihood and Bayesian marginalization analyses of the BOOMERANG cosmic microwave background experiment and supernovae (de Bernardis et al. 2002). The contours are from the de Bernardis et al. (2002) paper (reproduced here by permission) and are for the combined likelihood of the supernovae and the BOOMERANG constraints. The three lines we overplotted are the estimated 1-sigma constraints on Ω_M from cluster temperature function evolution.

5. Future work

One such survey is the DUET (Dark Universe Exploration Telescope) survey, a proposed MIDEX mission, principal investigator Rob Petre (GSFC). The proposed survey is of 10,000 square degrees, with a bandpass of 0.2–6.0 keV, effectively spanning most of a cluster's X-ray emission. Flux limits of 3×10^{-14} erg s^{-1} cm^{-2} goes sufficiently deep to detect all clusters and groups to $z < 0.4$ and all massive clusters with $kT > 3$ keV to $z \sim 1$ (thus including all clusters more massive than Coma ($kT \sim 6$ keV). The masses of the associated dark matter halos span 3 orders of magnitude. The total number of clusters in this survey would be between 10,000–50,000. One thousand of these clusters would have sufficient counts to obtain temperatures. Along with the clusters would be 100,000–200,000 AGN. 1000 cluster temperatures and 20,000 cluster luminosities is sufficient to place statistical and systematic constraints on Ω_M of $\pm 5\%$. Correspondingly, constraints on Ω_B are possible to $\pm 10\%$, σ_8 to $\pm 5\%$ and $n \pm 5\%$ (DUET team, private communication; see also Mohr et al. 2000).

It is important to note that while the statistical uncertainties with even the ASCA EMSS sample are nearly the same size as the systematic uncertainties, the future of reducing the total uncertainty with large numbers of clusters is not in reducing the statistical uncertainty, which can be reduced at large redshift. But at low redshift ($z < 0.1$), we are already very nearly volume-limited at $kT > 4$ keV. The future of this business is in reducing the *systematic* uncertainties, which can be reduced or at least better quantified by self-consistent studies of a large number of clusters. Systematics such as the evolution of the $L_x - T_x$ relation and the size of the dispersion in that relation or a reliable normalization of the $M_{virial} - T_x$ relation are necessary to improve the reliability of these techniques. It is important to note that pointed cluster observations by XMM and Chandra will also reduce some of these systematic uncertainties: the more we understand clusters the more we can reduce the effect of such systematics on the estimates of cosmological parameters.

The constraints on these and the "w" parameter (relevant to Λ and the equation of state of the vacuum energy density) are complementary to the constraint boundaries placed by the cosmic microwave background fluctuations and the Type Ia supernovae. Such a survey will also provide a sample suitable for the Sunyaev-Zel'dovich angular diameter test, which, in conjunction with other constraints on the Hubble constant, places yet another constraint on the geometry of the universe.

6. Conclusions

• Cluster studies have already produced robust and reliable estimates of Ω_M, and are capable of doing even better with an order of magnitude more clusters and the subsequent reduction of statistical *and* systematic uncertainties.

• The gas fraction estimates in clusters of $\sim 10\%$ imply $\Omega_M \sim 0.3$ in combination with big bang nucleosynthesis constraints on Ω_b. The gas fraction when corrected by the contribution from the mass associated with galaxies represents very nearly the baryonic fraction of clusters, unless there is significant amounts of intracluster cold dark baryonic material that is *not* associated with the cluster galaxies.

• Cluster evolution and cluster temperature functions are consistent with $\Omega_M \sim 0.3$ in a flat cosmology and somewhat higher in an open cosmology.

• Cluster cosmology constraints are complementary to those supplied by the cosmic microwave background and supernovae studies, and therefore play an important role in precision cosmology.

MD would like to acknowledge her collaborators and co-authors in this project, but special acknowledgments are due to Mark Voit, Isabella Gioia, and John Stocke for their contributions and support throughout. MD also acknowledges the financial support of a NASA LTSA (NAG-3257) and several NASA grants for data analysis of ASCA and ROSAT data, including NAG5-6236, NAG5-2615, and NAG5-2570. Chandra data analysis was made possible by grant GO0-1063A.

REFERENCES

ALLEN, S. W. 1998 *MNRAS*, **296**, 392.

ARNAUD, M., NEUMANN, D. M., AGHANIM, N., GASTAUD, R., MAJEROWICZ, S., & HUGHES, J. P. 2001 *A&A* **365** L80.

BÖHRINGER, H., TANAKA, Y., MUSHOTZKY, R. F., IKEBE, Y., & HATTORI, M. 1998 *A&A*, **334**, 789.

BOLDT, E., McDONALD, F. B., RIEGLER, G., & SERLEMITSOS, P. 1966 *Phys. Rev. Lett.* **17**, 447.

BRIEL, U. G., HENRY, J. P., & BÖHRINGER, H. 1992 *A&A* **259**, L31.

BRYAN, G. L. & VOIT, G. M. 2001 *ApJ* **556**, 590.

DAVID, L. P., ARNAUD, K. A., FORMAN, W., & JONES, C. 1990 *ApJ* **356**, 32.

DE BERNARDIS, P., ADE, P. A. R., BOCK, J. J., BOND, J. R., BORRILL, J., BOSCALERI, A., COBLE, K., CONTALDI, C. R., CRILL, B. P., DE TROIA, G., FARESE, P., GANGA, K., GIACOMETTI, M., HIVON, E., HRISTOV, V. V., IACOANGELI, A., JAFFE, A. H., JONES, W. C., LANGE, A. E., MARTINIS, L., MASI, S., MASON, P., MAUSKOPF, P. D., MELCHIORRI, A., MONTROY, T., NETTERFIELD, C. B., PASCALE, E., PIACENTINI, F., POGOSYAN, D., PO-

LENTA, G., PONGETTI, F., PRUNET, S., ROMEO, G., RUHL, J. E., SCARAMUZZI, F. 2002 *ApJ* **564**, 559; astro-ph/0105296.

BURLES, S. & TYTLER, D. 1998 *ApJ* **507**, 732.

DONAHUE, M., VOIT, G. M., GIOIA, I. M., LUPPINO, G., HUGHES, J. P., & STOCKE, J. T. 1998 *ApJ* **502**, 550.

DONAHUE, M., VOIT, G. M., SCHARF, C. A., GIOIA, I. M., MULLIS, C. R., HUGHES, J. P., STOCKE, J. T. 1999 *ApJ* **527**, 525.

DONAHUE, M. & VOIT, G. M. 1999 *ApJ* **523**, L137.

EDGE, A. C. & STEWARD, G. C. 1991 *MNRAS* **252**, 414.

EKE, V. R., COLE, S., FRENK, C. S., & HENRY, J. P. 1998 *MNRAS* **298**, 1145.

EVRARD, A. E. 1997 *MNRAS* **292**, 289.

EVRARD, A. E., VIRGO CONSORTIUM COLLABORATION 2001. In *2001 Moriond proceedings, Galaxy Clusters and the High Redshift Universe Observed in X-rays*, (ed. D. Neumann), cdrom.

FUKUGITA, M., HOGAN, C. J., & PEEBLES, P. J. E. 1998 *ApJ* **503**, 518.

GIOIA, I. M., MACCACARO, T., SCHILD, R. E., WOLTER, A., STOCKE, J. T., MORRIS, S. L., & HENRY, J. P. 1990 *ApJS* **72**, 567.

GIOIA, I. M., HENRY, J. P., MACCACARO, T., MORRIS, S. L., STOCKE, J. T., & WOLTER, A. 1990 *ApJ* **356**, L35.

GREGO, L., CARLSTROM, J. E., REESE, E. D., HOLDER, G. P., HOLZAPFEL, W. L., JOY, M. K., MOHR, J. J., & PATEL, S. 2001 *ApJ* **552**, 2.

GURSKY, H., KELLOGG, E., MURRAY, S., LEONG, C., TANANBAUM, H., & GIACCONI, R. 1971 *ApJ* **167**, L81.

HENRY, J. P., GIOIA, I. M., MACCACARO, T., MORRIS, S. L., STOCKE, J. T., & WOLTER, A. 1992 *ApJ* **386**, 408

HENRY, J. P. 2000 *ApJ* **534**, 565.

HENRY, J. P. & ARNAUD, K. A. 1991 *ApJ* **372**, 410.

HJORTH, J., OUKBIR, J., & VAN KAMPEN, E. 1998 *MNRAS* **298**, L1.

JELTEMA, T. E., CANIZARES, C. R., BAUTZ, M. W., MALM, M. R., DONAHUE, M., GARMIRE, G. P. 2001 *ApJ* **562**, 124; astro-ph/0107314.

LACEY, C. & COLE, S. 1993 *MNRAS* **262**, 627.

LILJE, P. B. 1992 *ApJ* **386**, L33.

MARKEVITCH, M. 1998 *ApJ* **504**, 27.

MCNAMARA, B. R., WISE, M., NULSEN, P. E. J., DAVID, L. P., SARAZIN, C. L., BAUTZ, M., MARKEVITCH, M., VIKHLININ, A., FORMAN, W. R., JONES, C., & HARRIS, D. E 1998 *ApJ* **534**, L135.

MOHR, J. J., MATHIESEN, B., & EVRARD, A. E. 1999 *ApJ* **517**, 627.

MOHR, J. J., HOFFMAN, M. B., BIALEK, J. J., & EVRARD, A. E. 2000 *AAS-HEAD*, **32**, 14.06.

OUKBIR, J. & BLANCHARD, A. 1992 *A&A* **262**, L21.

PRESS, W. H. & SCHECHTER, P. 1974 *ApJ* **187**, 425.

REESE, E. D., MOHR, J. J., CARLSTROM, J. E., JOY, M. K., GREGO, L., HOLDER, G. P., HOLZAPFEL, W. L., HUGHES, J. P., PATEL, S., & DONAHUE, M. 2001 *ApJ* **533**, 38.

ROUSSEL, H., SADAT, R., & BLANCHARD, A. 2000 *A&A* **361**, 429.

SOUCAIL, G., OTA, N., BÖHRINGER, H., CZOSKE, O., HATTORI, M., & MELLIER, Y. 2000 *A&A* **355** 433.

STOCKE, J. T., MORRIS, S. L., GIOIA, I. M., MACCACARO, T., SCHILD, R., WOLTER, A., FLEMING, THOMAS A., & HENRY, J. P. 1991 *ApJS*, **76**, 813.

TRAN, K.-V. H., KELSON, D. D., VAN DOKKUM, P., FRANX, M., ILLINGWORTH, G. D., & MAGEE, D. 1999 *ApJ* **522**, 39.

VAN DER MAREL, R. 1991 *MNRAS* **253**, 710.

WHITE, S. D. M., NAVARRO, J. F., EVRARD, A. E., & FRENK, C. S. 1996 *Nature* **366**, 429.

WU, X.-P., CHIUEH, T., FANG, L.-Z., & XUE, Y.-J. 1998 *MNRAS*, **301**, 861.

Tracking the baryon density from the Big Bang to the present

By GARY STEIGMAN

Departments of Physics and Astronomy, The Ohio State University, Columbus, OH 43210, USA

The primordial abundances of deuterium, helium, and lithium probe the baryon density of the universe only a few minutes after the Big Bang. Of these relics from the early universe, deuterium is the baryometer of choice. After reviewing the current observational status (a moving target!), the BBN baryon density is derived and compared to independent estimates of the baryon density several hundred thousand years after the Big Bang (as inferred from CMB observations) and at present, more than 10 billion years later. The excellent agreement among these values represents an impressive confirmation of the standard model of cosmology, justifying—indeed, demanding—more detailed quantitative scrutiny. To this end, the corresponding BBN-predicted abundances of helium and lithium are compared with observations to further test and constrain the standard, hot, big bang cosmological model.

1. Introduction

As progress is made towards a new, precision era of cosmology, *redundancy* will play an increasingly important role. As cosmology is an *observational* science, it will be crucial to avail ourselves of multiple, independent tests of, and constraints on, competing cosmological models and their parameters. Furthermore, such redundancy may provide the only window on systematic errors which can impede our progress or send us off in unprofitable directions. To illustrate the efficacy of such an approach in modern cosmology, I'll track the baryon density of the Universe as revealed early on (first few minutes) by Big Bang Nucleosynthesis (BBN), later (few hundred thousand years) as coded in the fluctuation spectrum of the Cosmic Microwave Background (CMB) radiation, and up to the present, approximately 10 Gyr after the expansion began. As theory suggests and terrestrial experiments confirm, baryon number should be preserved throughout these epochs in the evolution of the universe, so that the number of baryons ($\equiv$ nucleons) in a comoving volume *should* be unchanged from BBN to today. As a surrogate for identifying a comoving volume, we may compare the baryon/nucleon density to the density of CMB relic photons. Except for the additional photons produced when $e^{\pm}$ pairs annihilate, the number of photons in our comoving volume is also preserved. As a result, the baryon density may be tracked through the evolution of the universe utilizing the nucleon-to-photon ratio η where,

$$\eta \equiv n_{\mathrm{N}}/n_\gamma \ . \tag{1.1}$$

Since the temperature of the CMB fixes the present number density of relic photons, the fraction of the critical mass/energy density in baryons today ($\Omega_{\mathrm{B}} \equiv \rho_{\mathrm{B}}/\rho_{crit}$) is directly related to η,

$$\eta_{10} \equiv 10^{10}\eta = 274\Omega_{\mathrm{B}}h^2 \ , \tag{1.2}$$

where the Hubble parameter, measuring the present universal expansion rate, is $\mathrm{H}_0 \equiv 100h$ km s^{-1} Mpc^{-1}. According to the *HST* Key Project, $h = 0.72 \pm 0.08$ (Freedman et al. 2001).

For several decades now the best constraints on η have come from the comparison of the predictions of BBN with the observational data relevant to inferring the primordial abundances of the relic nuclides D, ^{3}He, ^{4}He, and ^{7}Li. For recent reviews, the interested reader is referred to Olive, Steigman, & Walker (2000), Burles, Nollett, & Turner (2001), Cyburt, Fields, & Olive (2001), and references therein. However, the time is rapidly approaching when new data of similar accuracy will be available to probe the baryon density at later epochs. Indeed, recent CMB data has very nearly achieved this goal. This same level of precision is currently lacking for the present universe estimates but, such as they are, they do permit us to compare and contrast independent estimates of η (or $\Omega_B h^2$) at three widely separated eras in the evolution of the universe. In the next section the BBN bounds on the nucleon density are presented and their implications for the dark baryon and dark matter problems described. Next, we turn to the CMB estimates of the baryon density, comparing them to those from BBN. Last (but not least), we turn to an estimate of the present universe baryon density utilizing data whose interpretation avoids the need to adopt a relation between mass and light in the universe.

In recent years cosmological research has been data-driven and we have become used to looking forward with great anticipation to the latest observational results, and greedily wishing for more. Be careful of what you wish for! Not all the new data has led us unerringly along the right path. Indeed, within a few weeks of this meeting new data became available on the CMB angular fluctuation spectrum (Halverson et al. 2001; Netterfield et al. 2001; Lee et al. 2001) and on the deuterium abundance (Levshakov, Dessauges-Zavadsky, D'Odorico, & Molaro 2001; Pettini & Bowen 2001) which change dramatically some of the results I presented in my talk at this Symposium. In the interest of preserving some of the historical record, I will comment on the older data and their implications, while reserving the most recent observations for my conclusions regarding the baryon density of the universe.

2. The baryon density during the first few minutes

During the first few minutes in the evolution of the universe the density and temperature were sufficiently high for nuclear reactions to occur in the time available. As the universe expanded, cooled, and became more dilute, the "rolling blackout" became permanent and the universal nuclear reactor went offline. The abundances of the light nuclei formed during this limited epoch of primordial alchemy were determined, for the most part, by the competition between the time available (the universal expansion rate) and the density of the reactants: nucleons (neutrons and protons). The abundances predicted for nucleosynthesis in the standard cosmological model (SBBN) are shown (as a function of η) in Figure 1.

The abundances of D, ^{3}He, and ^{7}Li are "rate limited," being determined by the competition between the nuclear production/destruction rates and the universal expansion rate. As such, they are sensitive to the nucleon density and have the potential to serve as "baryometers." In contrast, the nuclear reactions which build ^{4}He are so rapid and, because its destruction is inhibited (due to its strong binding and the lack of a stable nuclide at mass-5), the helium-4 mass fraction, Y, is insensitive to η. Since virtually all the neutrons available at BBN are incorporated into ^{4}He, Y is determined mainly by the neutron-to-proton ratio which is controlled by the competition between the weak interaction rate ("β-decay") and the expansion rate. As a result, since the neutron lifetime is known very accurately, the ^{4}He abundance is a sensitive probe of the early universe expansion rate.

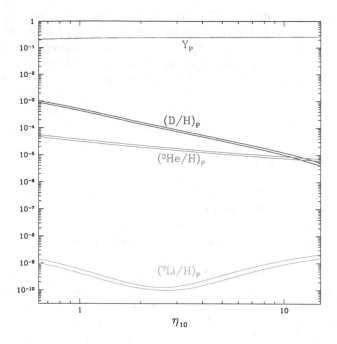

FIGURE 1. The SBBN-predicted abundances of D, ^{3}He, ^{7}Li (by number relative to hydrogen) and the ^{4}He mass fraction Y, as a function of the nucleon-to-photon ratio η. The widths of the bands reflect the BBN uncertainties associated with the nuclear and weak interaction rates.

2.1. *Deuterium—the baryometer of choice*

Of the three relic nuclides whose primordial abundances may be probes of the baryon density (D, ^{3}He, ^{7}Li), deuterium is the baryometer of choice. First and foremost, the predicted primordial abundance has a significant dependence on the nucleon density (D/H $\propto \eta^{-1.6}$). As a result, if the primordial abundance is known to, say, 10%, then the baryon density (η) can be determined to $\sim 6\%$; truly precision cosmology! Equally important, as Epstein, Lattimer, & Schramm (1976) showed long ago, BBN is the only astrophysical site where an "interesting" abundance of deuterium may be produced (D/H $\gtrsim 10^{-5}$); the relic abundance is not enhanced by post-BBN production. Furthermore, as primordial gas is cycled through stars, deuterium is completely destroyed (because of the small binding energy of the deuteron, destruction occurs during the pre-main sequence evolution, when the stars are fully mixed). As a result, the abundance of deuterium has only decreased (or, remained close to its primordial value) since BBN.

$$(D/H)_{\text{NOW}} \leqslant (D/H)_{\text{BBN}} \quad , \tag{2.1}$$

where "NOW" refers to the "true" deuterium abundance in any system observed at any time (e.g. the solar system; the interstellar medium; high redshift, low metallicity QSO absorbers, etc.).

There is, however, one potentially serious problem associated with using deuterium as a baryometer. The atomic spectra of deuterium and hydrogen are identical, save for the isotope shift associated with the deuteron-proton mass difference. As a result, a small amount of hydrogen at the "wrong" velocity (an "interloper") can masquerade as deuterium, so that

$$(D/H)_{\text{NOW}} \leqslant (D/H)_{\text{OBS}} \quad . \tag{2.2}$$

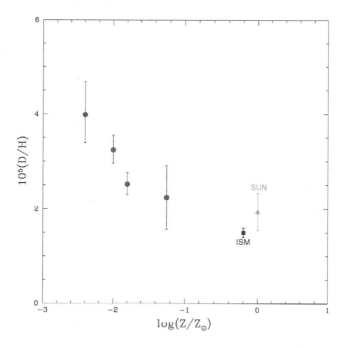

FIGURE 2. The deuterium abundance by number relative to hydrogen derived from observations of high-z, low-Z QSO absorption line systems as a function of metallicity (relative to solar). Also shown are the deuterium abundances inferred for the local interstellar medium (Linsky & Wood 2000) and for the presolar nebula (Gloeckler & Geiss 2000).

A comparison of equations 2.1 and 2.2 reveals the problem: how to relate $(D/H)_{OBS}$ to $(D/H)_{BBN}$? One approach is to concentrate on observing deuterium in those high redshift (high-z), low metallicity (low-Z) systems where not much gas will have been cycled through stars and the deuterium ("NOW") should be essentially primordial (BBN). If so, there should be no variation (outside of the statistical errors) with metallicity. To test for interlopers the best approach (the favorite of theorists!) is to have lots of data. Observers are making great progress towards this goal, but the road has not always been straight.

After some false starts, there were four high-z, low-Z QSO absorption line systems where deuterium had been detected as of the time of this Symposium in April 2001 (Burles & Tytler 1998a,b; O'Meara et al. 2001; D'Odorico, Dessauges-Zavadsky, & Molaro 2001). A fifth system (Webb et al. 1997), with possibly a much higher D/H, is widely agreed to have insufficient velocity data to rule out contamination by a hydrogen interloper (Kirkman et al. 2001). However, these data on D/H, displayed as a function of metallicity in Figure 2, appear to challenge our expectations (of a low-metallicity, deuterium "plateau") in that they suggest an anticorrelation between D/H and metallicity. Has deuterium actually been destroyed since BBN, and is the BBN abundance at least as high as the highest data point? This would be most surprising in such low metallicity systems (see, e.g. Jedamzik & Fuller 1997). Perhaps one or more of the "high" points is "too high" because of an interloper. To test for this possibility, let's display the same data as a function of the hydrogen column density (O'Meara et al. 2001); see Figure 3. It might be expected that the lower hydrogen column density absorbers could be more easily contaminated. Two of the systems have relatively low H-column densi-

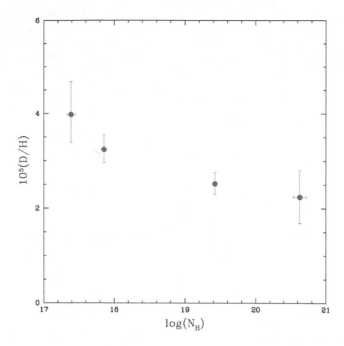

FIGURE 3. The same deuterium abundances as in Figure 2 plotted as a function of the hydrogen column density N_H (cm^{-3}) in the absorbing cloud.

ties ($\lesssim 10^{18}$cm^{-3} ≡ Lyman Limit Systems (LLS)) and two have considerably higher N_H ($\gtrsim 10^{19}$cm^{-3} ≡ Damped Lyman Alpha (DLA) absorbers).

Indeed, the hint from Figure 3 is that interlopers may be playing a role in either or both of the LLS. If so, perhaps we should identify only the DLAs with the primordial deuterium abundance. However, given the small number of data points, this would be premature. At the Symposium I adopted a weighted average ($\langle D/H \rangle = 2.9 \pm 0.3 \times 10^{-5}$) and called for more data. Observers weren't long in responding. But, I hadn't anticipated what they would find.

Further observations by Levshakov, Dessauges-Zavadsky, D'Odorico, & Molaro (2001) of the D'Odorico, Dessauges-Zavadsky, & Molaro (2001) absorber (the low D/H point at the highest N_H in Fig. 3) revealed a more complex velocity structure and led to an upward revision in the derived D/H (which, by the way, is by more than the previous statistical error estimate, reinforcing the potential for systematic errors to wreak havoc). So far, so good, now that the four D/H determinations are more consistent with one another, alleviating the need to invoke interloper contamination. However, in the meanwhile Pettini & Bowen (2001) weighed in with a new deuterium detection in a DLA. The Pettini-Bowen abundance is smaller than all the other determinations, indeed smaller than (although within the errors of) the presolar nebula abundance of Gloeckler & Geiss (2000). The current data (as of July 2001) are shown in Figure 4.

What to do? The dispersion among these four determinations hints that one or more may be wrong. Until this puzzle is resolved by more data (the last refuge of the theorist), I will adopt as a default estimate the one derived by O'Meara et al. (2001): $(D/H)_{BBN} \equiv 3.0 \pm 0.4 \times 10^{-5}$. The consequences for the BBN derived baryon density are shown in Figure 5 where the overlap between the predicted and "observed" primordial abundances are used to constrain η.

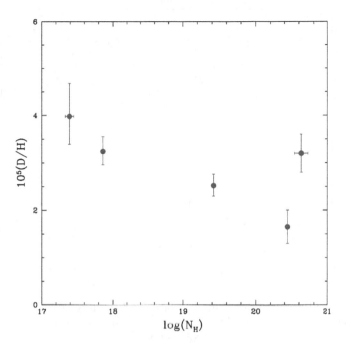

FIGURE 4. The current (July 2001) deuterium abundances plotted as a function of the hydrogen column density N_H (cm^{-3}) in the absorbing cloud. See the text for references.

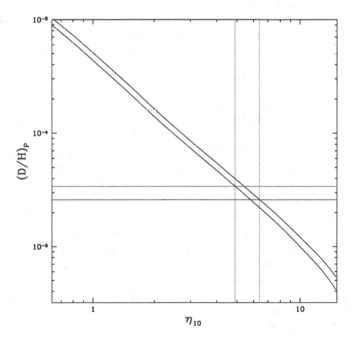

FIGURE 5. The band stretching from upper left to lower right is the BBN-predicted deuterium abundance (as in Fig. 1). The horizontal band is the observational estimate of the primordial abundance (see the text). The vertical band provides an estimate of the BBN-derived baryon density.

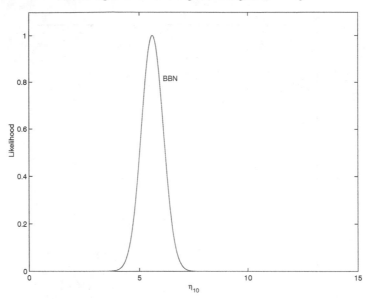

FIGURE 6. The likelihood distribution (normalized to unity at maximum) for the baryon-to-photon ratio derived from BBN and the (adopted) primordial abundance of deuterium.

2.2. *The BBN baryon density*

From a careful comparison between the BBN predicted abundance (including errors—which are subdominant) and the adopted primordial value, the baryon density when the universe is less than a half hour old may be constrained (see Figure 6). For cosmology, this is truly a "precision" determination. Whether it is accurate, only time will tell.

$$\eta_{10}(\mathrm{BBN}) = 5.6 \pm 0.5 \quad (\Omega_{\mathrm{B}}h^2 = 0.020 \pm 0.002) \quad . \qquad (2.3)$$

This range for the baryon density poses some interesting challenges to our view of the universe. These challenges may be seen by comparing the BBN estimate of Ω_{B} with those found by adding up all the baryons associated with the "luminous" material observed in the present/recent universe at $z \lesssim 1$ (Persic & Salucci 1992), and with estimates of the total mass density at present. These comparisons are shown in Figure 7 where the various density parameter estimates/ranges are plotted as a function of the Hubble parameter H_0 (recall that, according to Freedman et al. (2001), $H_0 = 72 \pm 8$).

The gap between the upper limit to luminous baryons and the BBN band is the "dark baryon problem": not all the baryons expected from BBN have been seen in the present universe. Perhaps it is hubris to expect that all baryons will choose to radiate (or absorb) in those parts of the spectrum we can see, or which our instruments can record. Indeed, from the absorption observed in the Lyman-alpha forest at redshifts of order 2–3 (Weinberg, Miralda-Escudé, Hernquist, & Katz 1997), it seems clear that the density of baryons in the present universe is much larger than the upper bound to luminous baryons shown in Figure 7. We return to a different estimate of the baryon density in the present/recent universe in § 4.

The gap between the BBN band and the band labeled Ω_{M} is one aspect of the classical dark matter problem: the mass density inferred from the motions of galaxies exceeds the baryon density derived from BBN. Aside from those solutions which resort to modifying gravity (e.g. Sanders (2001) this volume), the standard assumption is that this gap

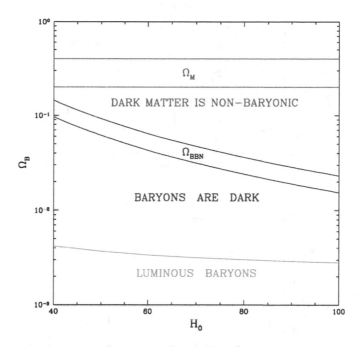

FIGURE 7. Several estimates of the baryon density (relative to the critical density) in the present universe as a function of the Hubble parameter H_0. The band labeled "BBN" is the early universe baryon density estimate (see eq. 2.3). The line labeled "Luminous Baryons" is the upper bound to estimates of the density of baryons seen at present in emission or absorption (Persic & Salucci 1992). The band labeled Ω_M (= 0.3 ± 0.1) is an estimate of the total mass density in nonrelativistic particles at present.

provides evidence for the dominance at present of non-baryonic dark matter. Much of this Symposium was devoted to discussions of possible dark matter candidates and the means for their detection.

3. The baryon density at a few hundred thousand years

The early universe is radiation dominated. As the universe expands and cools, the density in non-relativistic matter becomes relatively more important, eventually dominating after a few hundred thousand years. At this stage perturbations can begin to grow under the influence of gravity and, on scales determined by the relative density of baryons, oscillations in the baryon-photon fluid will develop. When, at a redshift ~ 1100, the electron-proton plasma combines to form neutral hydrogen, the CMB photons are freed to propagate throughout the universe. These CMB photons preserve the record of the baryon-photon oscillations through very small temperature fluctuations in the CMB spectrum which have been detected by the newest generation of experiments, beginning with COBE (Bennett et al. 1996) and continuing with the exciting BOOMERANG (de Bernardis et al. 2000; Lange et al. 2000) and MAXIMA-1 (Hanany et al. 2000) results which appeared somewhat more than a year ago. In Figure 8 we illustrate the status quo ante with the dramatic and challenging results from those earlier data.

The relative heights of the odd and even peaks in the CMB angular fluctuation spectrum depend on the baryon density and the early BOOMERANG and MAXIMA-1 data favored a "high" baryon density (compare the "BBN case", $\eta_{10} = 5.6$, in the upper

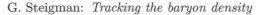

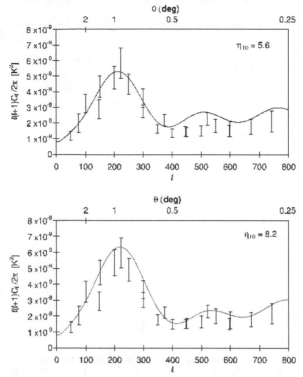

FIGURE 8. The CMB angular fluctuation spectra for two models which differ only in the adopted baryon density. The BBN inferred baryon density is shown in the upper panel and, for comparison, a higher baryon density model is shown in the lower panel. The data are from the "old" BOOMERANG and MAXIMA-1 observations; see the text for references.

panel with that for a baryon density some 50% higher shown in the lower panel). At the time of this Symposium, these were the extant data and they posed a challenge to the consistency of the standard model of cosmology.

At the Symposium we were told that new data would be forthcoming shortly and, the observers didn't disappoint. In less than a month new (and some revised) data appeared (Halverson et al. 2001; Netterfield et al. 2001; Lee et al. 2001) which have eliminated the challenge posed by the older data. Although the extraction of cosmological parameters from the CMB data can be very dependent on the priors adopted in the analyses (see Kneller et al. 2001), the inferred baryon density is robust. Kneller, Scherrer, Steigman, & Walker (2001) find,

$$\eta_{10}(\text{CMB}) = 6.0 \pm 0.6 \quad (\Omega_B h^2 = 0.022 \pm 0.002) \quad . \tag{3.1}$$

In Figure 9 is shown the comparison of the likelihoods for the nucleon-to-photon ratio from BBN, when the universe was some tens of minutes old, and from the CMB, some few hundred thousand years later. The excellent agreement between the two, independent estimates is a spectacular success for the standard model of cosmology and an illustration of the great potential for precision tests of cosmology.

4. The Baryon Density At 10 Gyr

As already discussed in §2 (see Figure 7), the amount of baryons "visible" in the present universe is small compared to that expected on the basis of BBN (and, as seen

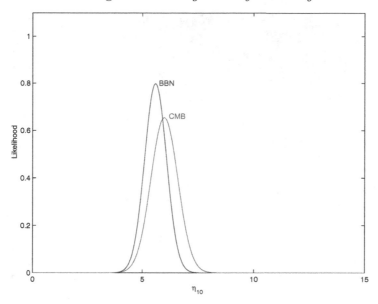

FIGURE 9. The likelihood distributions (normalized to equal areas under the curves) for the baryon-to-photon ratios derived from BBN and from the CMB.

in § 3, to that revealed by the CMB data). If most baryons in the present universe are dark, how can their density be constrained? There are a variety of approaches. Many depend on assumptions concerning the relation between mass and light, or require the adoption of a specific model for the growth of structure. In the approach utilized here we attempt to avoid such model-dependent assumptions. Instead, we use the data from the SNIa magnitude-redshift surveys (Perlmutter et al. 1997; Schmidt et al. 1998; Perlmutter et al. 1999), along with the *assumption* of a flat universe (which receives strong support from the newest CMB data; Halverson et al. 2001; Netterfield et al. 2001; Lee et al. 2001) to pin down the total matter density (Ω_M), which will be combined with an estimate of the universal baryon fraction ($f_B \equiv \Omega_B/\Omega_M$) derived from studying the X-ray emission from clusters of galaxies. For more details on this approach, see Steigman, Hata, & Felten (1999) and Steigman, Walker, & Zentner (2000).

In Figure 10 are shown the SNIa-constrained 68% and 95% contours in the $\Omega_\Lambda - \Omega_M$ plane. The expansion of the universe is currently accelerating for those models which lie above the (dashed) $q_0 = 0$ line. The $k = 0$ line is for a "flat" (zero 3-space curvature) universe. As shown in Steigman, Walker, & Zentner (2000), adopting the assumption of flatness and assuming the validity of the SNIa data, leads to a reasonably accurate ($\sim 25\%$) estimate of the present matter density.

$$\Omega_M(\text{SNIa; Flat}) = 0.28^{+0.08}_{-0.07} \ . \qquad (4.1)$$

As the largest collapsed objects, rich clusters of galaxies provide an ideal probe of the baryon *fraction* in the present universe. Observations of the X-ray emission from clusters of galaxies permit constraints on the hot gas content of such clusters which, when corrected for baryons in stars (but, not for any dark baryons!), may be used to estimate f_B. From observations of the Sunyaev-Zeldovich effect in X-ray clusters, Grego et al. (2001)constrain the hot gas fraction which Steigman, Kneller, & Zentner (2001)

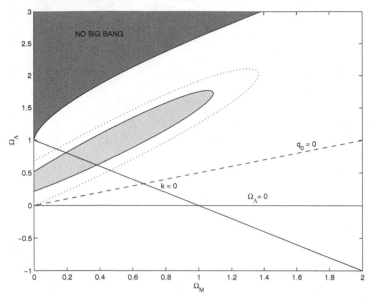

FIGURE 10. The 68% (solid) and 95% (dotted) contours in the $\Omega_\Lambda - \Omega_M$ plane allowed by the SNIa magnitude-redshift data (see the text for references). Geometrically flat models lie along the diagonal line labeled $k = 0$.

have used to estimate f_B and to derive a present-universe ($t_0 \approx 10$ Gyr; $z \lesssim 1$) baryon density.

$$\eta_{10}(\text{SNIa; Flat}) - 5.1^{+1.8}_{-1.4} \quad (\Omega_B h^2 = 0.019^{+0.007}_{-0.005}) \quad . \tag{4.2}$$

In Figure 11 the corresponding likelihood distribution for the present universe baryon density is shown ("SNIa") along with those derived earlier from deuterium and BBN, and from the CMB fluctuation spectra. Although the uncertainties are largest for this present-universe value, it is in excellent agreement with the other, independent estimates.

5. Summary and discussion

The abundances of the relic nuclides produced during BBN encode the baryon density during the first few minutes in the evolution of the universe. Of these relics from the early universe, deuterium is the barometer of choice. The deuterium abundance in relatively unprocessed material, such as the high-z, low-Z QSO absorption line systems should be very nearly primordial. At present there are data for five such systems. Although the statistical accuracies of these data are high, the dispersion among them in the derived D/H ratio is surprisingly large, suggesting that systematic errors (interlopers?, complex velocity structure?) may effect one or more of these determinations. Nonetheless, these data seem consistent with a primordial abundance $(D/H)_P = 3.0 \pm 0.4 \times 10^{-5}$ (O'Meara et al. 2001) which was adopted in §2 to derive $\eta_{10}(\text{BBN}) = 5.6 \pm 0.5$ ($\Omega_B h^2 = 0.020 \pm 0.002$).

Several hundred thousand years later, when the universe became transparent to the CMB radiation, the baryon density was imprinted on the temperature fluctuation spectrum which has been observed by the COBE (Bennett et al. 1996), BOOMERANG (Netterfield et al. 2001), MAXIMA (Lee et al. 2001), and DASI (Halverson et al. 2001) experiments. For determining the baryon density, these current CMB data have a precision approaching that of BBN: $\eta_{10}(\text{CMB}) = 6.0 \pm 0.6$ ($\Omega_B h^2 = 0.022 \pm 0.002$). The

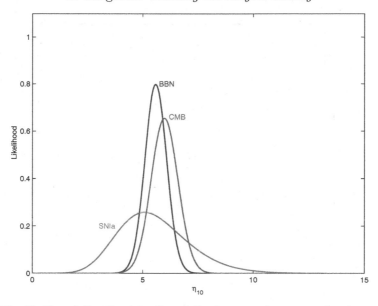

FIGURE 11. The likelihood distributions (normalized to equal areas under the curves) for the baryon-to-photon ratios derived from BBN, from the CMB, and for the present universe (SNIa) using the SNIa and X-ray cluster data, and the assumption of a flat universe.

excellent agreement between the BBN and CMB values (see Fig. 9) provides strong support for the standard model of cosmology.

In the present universe most baryons are dark (η(LUM) $\ll$ η(BBN) $\approx$ η(CMB)), so that estimates of the baryon density some 10 billion years after the expansion began are more uncertain and, often model-dependent. In §4 we combined an estimate of the total matter density (Ω_M) derived from the SNIa magnitude-redshift data (Perlmutter et al. 1997; Schmidt et al. 1998; Perlmutter et al. 1999) and the assumption of a flat universe ($\Omega_M = 0.28^{+0.08}_{-0.07}$), with the universal baryon fraction inferred from X-ray observations of clusters of galaxies (Grego et al. 2001) to derive η_{10}(SNIa; Flat) $= 5.1^{+1.8}_{-1.4}$ ($\Omega_B h^2 = 0.019^{+0.007}_{-0.005}$) (Steigman, Kneller, & Zentner 2000). Although of much lower statistical accuracy, this estimate of the present universe baryon density is in complete agreement with those from BBN and the CMB (see Fig. 11). I note in passing that if the mass of dark baryons in clusters is similar to the stellar mass, this present-universe baryon density estimate would increase by $\sim$ 10%, bringing it into essentially perfect overlap with the BBN and CMB values.

The concordance of the standard, hot, big bang cosmological model is revealed clearly by the overlapping likelihood distributions for the universal density of baryons shown in Fig. 11. As satisfying as this agreement might be, it should impel us to action, not complacency. How may we test further the standard model? One way is to return to the relic nuclides which, so far, have been set aside: ^{4}He and ^{7}Li.

Consistency among the three, independent baryon density estimates permits us to identify a "best" value: $(\eta_{10})_{\text{best}} = 5.7$ (($\Omega_B h^2)_{\text{best}} = 0.021$). For this baryon-to-photon ratio the BBN-predicted primordial abundances are: $(D/H)_P = 2.9 \times 10^{-5}$, $Y_P = 0.248$, and $(Li/H)_P = 3.8 \times 10^{-10}$. These "best" estimates are shown by the "stars" in Figures 12 and 13. In those figures the bands reflect the BBN-predicted ^{4}He vs. D and ^{7}Li vs. D relations, including the nuclear and weak interaction physics uncertainties. It is clear

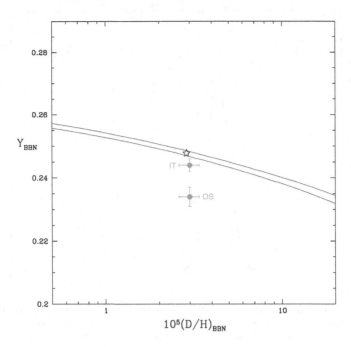

FIGURE 12. The BBN-predicted ^{4}He mass fraction (Y_{BBN}) versus the deuterium abundance (by number with respect to hydrogen: $(D/H)_{BBN}$) is shown by the band from left-to-right. The star corresponds to Y_{BBN} and $(D/H)_{BBN}$ for the "best" value of the universal density of baryons (see the text). The data points are shown at the O'Meara et al. (2001) deuterium abundance estimate and the IT and OS values for the helium abundance (see the text).

that the deuterium abundance is in excellent agreement with current data. What of ^{4}He and ^{7}Li?

At present there are two estimates for the primordial abundance of ^{4}He based on large (nearly) independent data sets and analyses of low-metallicity, extragalactic H II regions; see Fig. 12. The "IT" (Izotov, Thuan, & Lipovetsky 1994, Izotov & Thuan 1998) estimate of $Y_P(IT) = 0.244 \pm 0.002$ is only 2σ away from our best fit value of 0.248, while the "OS" determination (Olive & Steigman 1995, Olive, Skillman, & Steigman 1997, Fields & Olive 1998) of $Y_P(OS) = 0.234 \pm 0.003$ is nearly 5σ lower. Although it may be tempting to dismiss the OS estimate, recent high quality observations of a relatively metal-rich H II region in the SMC by Peimbert, Peimbert, & Ruiz (2000) reveal an abundance ($Y_{SMC} = 0.2405 \pm 0.0018$) which is *lower* than the IT primordial value. When this abundance is extrapolated to zero metallicity, Peimbert, Peimbert, & Ruiz (2000) find $Y_P(PPR) = 0.2345 \pm 0.0026$, in excellent—albeit accidental!—agreement with the OS value. The comparisons among different observations and between theory and observations suggest that unaccounted systematic errors (underlying stellar absorption weakening the helium emission lines?) may have contaminated at least some of the data.

The comparison between theory and data for ^{7}Li poses much more of a challenge; see Fig. 13. The data point plotted in Fig. 13, from Ryan et al. (2000): $(Li/H)_P = 1.23^{+0.68}_{-0.32} \times 10^{-10}$, is lower by a factor of three (~ 0.5 dex) than the standard model prediction (the "star" in Fig. 13). There are, however, good reasons to believe that the low value of the lithium abundance derived by Ryan et al. (2000) from observations of metal-poor, *old* stars (Ryan, Norris & Beers 1999) is not representative of the lithium

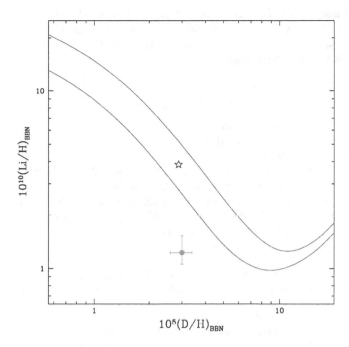

FIGURE 13. As in Fig. 12, but for lithium versus deuterium. The star is the BBN-predicted "best" value for ^{7}Li and D, while the ^{7}Li point is from Ryan et al. (2000).

abundance in the gas out of which these stars formed (Pinsonneault et al. 1999; Salaris & Weiss 2001; Pinsonneault et al. 2001; Theado & Vauclair 2001). Using a model for lithium depletion via rotational mixing over the long lifetimes of these stars, Pinsonneault et al. (2001) reanalyzed the Ryan, Norris, & Beers (1999) data and concluded that the primordial lithium abundance should be higher than the Ryan et al. (2000) value by $\approx 0.3 \pm 0.2$ dex, bringing the abundance inferred from the earliest generation of stars in the Galaxy in closer agreement with that expected for the "best" estimate of the baryon density in the standard cosmological model.

6. Conclusions

Increasingly precise observational data permit us to track the baryon density from the big bang to the present. At widely separated epochs from the first few minutes, through the first few hundred thousand years, to the present universe, a consistent value emerges, accurate to better than 10%: $\eta_{10} \approx 5.7$ ($\Omega_B h^2 \approx 0.021$). Such a low baryon density ($\Omega_B \approx 0.04$) reinforces the need for non-baryonic dark matter ($\Omega_B \lesssim \Omega_M/7$), which itself appears to be subdominant at present to an unknown form of dark energy. Precision cosmology has led us to an extreme form of the Copernican Principle! While this baryon density is fully consistent with present estimates of the primordial deuterium abundance, it challenges the data and "astrophysics" used to determine the primordial abundances of ^{4}He and ^{7}Li. If unresolved by new data, or better astrophysics, these conflicts might be providing a peek at new physics beyond the standard models of cosmology and/or particle physics. These are surely exciting times.

For valuable advice, assistance, and tutorials I wish to thank Gus Evrard, Jim Felten, Jim Kneller, Jeff Linsky, Joe Mohr, Paulo Molaro, Keith Olive, Manuel Peimbert, Max

Pettini, Marc Pinsonneault, Bob Scherrer, Evan Skillman, David Tytler, Sueli Viegas, Terry Walker, and Andrew Zentner. Praise is due Mario Livio and the efficient staff of the ST ScI for the smooth organization of an exciting meeting. My research is supported by DOE grant DE-FG02-91ER-40690.

REFERENCES

BENNETT, C. L., ET AL. 1996 *ApJ* **464**, L1.

BURLES, S. & TYTLER, D. 1998a *ApJ* **499**, 699.

BURLES, S. & TYTLER, D. 1998b *ApJ* **507**, 732.

BURLES, S., NOLLETT, K. M., & TURNER, M. S. 2001 *Phys. Rev.* **D63**, 063512.

CYBURT, R. H., FIELDS, B. D., & OLIVE, K. A. 2001 preprint; (astro-ph/0102179).

D'ODORICO, S., DESSAUGES-ZAVADSKY, M., & MOLARO, P. 2001 *A&A* **338**, L1.

EPSTEIN, R., LATTIMER, J., & SCHRAMM, D. N. 1976 *Nature* **263**, 198.

FIELDS, B. D. & OLIVE, K. A. 1996 *ApJ* **506**, 177.

FREEDMAN, W. L., ET AL. 2001 *ApJ* **553**, 47.

JEDAMZIK, K. & FULLER, G. 1997 *ApJ* **483**, 560.

GLOECKLER, G. & GEISS, J. 2000. In *Proceedings of IAU Symposium 198, The Light Elements and Their Evolution* (eds. L. da Silva, M. Spite, & J. R. Medeiros). p. 224. ASP Conference Series.

GREGO, L., ET AL. 2001 *ApJ* **552**, 2.

HALVERSON, N. W., ET AL. 2001 preprint; astro-ph/0103305.

IZOTOV, Y. I., THUAN, T. X., & LIPOVETSKY, V. A. 1994 *ApJ* **435**, 647.

IZOTOV, Y. I. & THUAN, T. X. 1998 *ApJ* **500**, 188.

KIRKMAN, D., ET AL. 2001 preprint; astro-ph/0104489.

KNELLER, J. P., SCHERRER, R. J., STEIGMAN, G., & WALKER, T. P. 2001 preprint; astro-ph/0101386.

LEE, A. T., ET AL. 2001 preprint; astro-ph/0104459.

LEVSHAKOV, S. A., DESSAUGES-ZAVADSKY, M., D'ODORICO, S., & MOLARO, P. 2001 preprint; astro-ph/0105529.

LINSKY, J. L. & WOOD, B. E. 2000 In *Proceedings of IAU Symposium 198, The Light Elements and Their Evolution* (eds. L. da Silva, M. Spite, & J. R. Medeiros). p. 141. ASP Conference Series.

NETTERFIELD, C. B., ET AL. 2001 preprint; astro-ph/0104460.

OLIVE, K. A. & STEIGMAN, G. 1995 *ApJS* **97**, 49.

OLIVE, K. A., SKILLMAN, E., & STEIGMAN, G. 1997 *ApJ* **483**, 788.

OLIVE, K. A., STEIGMAN, G., & WALKER, T. P. 2000 *Phys. Rep.* **333**. 389.

O'MEARA, J. M., ET AL. 2001 *ApJ* **552**, 718.

PEIMBERT, M., PEIMBERT, A., & RUIZ, M. T. 2000 *ApJ* **541**, 688.

PERLMUTTER, S., ET AL. 1997 *ApJ* **483**, 565.

PERLMUTTER, S., ET AL. 1999 *ApJ* **517**, 565.

PERSIC, M. & SALUCCI, P. 1992 *MNRAS* **258**, 14P.

PETTINI, M. & BOWEN, D. V. 2001 preprint; astro-ph/0104474.

PINSONNEAULT, M. H., WALKER, T. P., STEIGMAN, G., & NARAYANAN, V. K. 1999 *ApJ* **527**, 180.

PINSONNEAULT, M. H., STEIGMAN, G., WALKER, T. P., & NARAYANAN, V. K. 2001 preprint; astro-ph/0105439.

RYAN, S. G., NORRIS, J. E., & BEERS, T. C. 1999 *ApJ* **523**, 654.

RYAN, S. G., BEERS, T. C., OLIVE, K. A., FIELDS, B. D., & NORRIS, J. E. 2000 *ApJ* **530**, L57.

SALARAIS, M. & WEISS, A. 2001 preprint; astro-ph/0104406.

SANDERS, R. H. 2001 preprint; astro-ph/0106558.

SCHMIDT, B. P., ET AL. 1998 *ApJ* **507**, 46.

STEIGMAN, G. & FELTEN, J. E. 1995 *Spa. Sci. Rev.* **74**, 245.

STEIGMAN, G., HATA, N., & FELTEN, J. E. 1999 *ApJ* **510**, 564.

STEIGMAN, G., KNELLER, J. P., & ZENTNER, A. 2001 preprint; astro-ph/0102152.

STEIGMAN, G., WALKER, T. P., & ZENTNER, A. 2000 preprint; astro-ph/0012149.

THEADO, S. & VAUCLAIR, S. 2001 preprint; astro-ph/0106080.

WEBB, J. K., CARSWELL, R. F., LANZETTA, K. M., FERLET, R., LEMOINE, M., VIDAL-MADJAR, A., & BOWEN, D. V. 1997 *Nature* **388**, 250.

WEINBERG, D. H., MIRALDA-ESCUDÉ, J., HERNQUIST, L., & KATZ, N. 1997 *ApJ* **490**, 564.

Modified Newtonian Dynamics and its implications

By R. H. SANDERS

Kapteyn Astronomical Institute, Groningen, The Netherlands

Milgrom has proposed that the appearance of discrepancies between the Newtonian dynamical mass and the directly observable mass in astronomical systems could be due to a breakdown of Newtonian dynamics in the limit of low accelerations rather than the presence of unseen matter. Milgrom's hypothesis, modified Newtonian dynamics or MOND, has been remarkably successful in explaining systematic properties of spiral and elliptical galaxies and predicting in detail the observed rotation curves of spiral galaxies with only one additional parameter—a critical acceleration which is on the order of the cosmologically interesting value of cH_o. Here I review the empirical successes of this idea and discuss its possible extention to cosmology and structure formation.

---

1. Introduction

Modified Newtonian dynamics (MOND) is an *ad hoc* modification of Newton's law of gravity or inertia proposed by Milgrom (1983) as an alternative to cosmic dark matter. The motivation for this and other such proposals is obvious: So long as the only evidence for dark matter is its global gravitational effect, then its presumed exitance is not independent of the assumed form of the law of gravity or inertia on astronomical scales. In other words, either the universe contains large quantities of unseen matter, or gravity (or the response of particles to gravity) is not generally the same as it appears to be in the solar system.

The phenomenological foundations for MOND really come down to two observational facts about spiral galaxies: 1) The rotation curves of spiral galaxies are asymptotically flat, and 2) There is a well-defined relationship between the rotation velocity in spiral galaxies and the luminosity—the Tully-Fisher (TF) law (Tully & Fisher 1977). This latter implies a mass-velocity relationship of the form $M \propto V^\alpha$ where α is in the neighborhood of 4.

If one wants to modify gravity in some way to explain flat rotation curves or the exitance of a mass-rotation velocity relation for spiral galaxies, an obvious first choice would be to propose that gravitational attraction becomes more like 1/r beyond some length scale which is comparable to the scale of galaxies. So the modified law of attraction about a point mass M would read

$$F = \frac{GM}{r^2} f(r/r_o) \quad , \tag{1.1}$$

where r_o is a new constant of length with dimensions of a few kpc, and $f(x)$ is a function with the asymptotic behavior: $f(x) = 1$ where $x \ll 1$ and $f(x) = x$ where $x \gg 1$. Equating the centripetal to the gravitational acceleration in the limit $r \gg r_o$ would lead to a mass-asymptotic rotation velocity relation of the form $v^2 = GM/r_o$. This is true of any modification attached to a length scale. Milgrom realized that this was incompatible with the observed TF law unless, of course, the mass-to-light ratio (M/L) of the stellar population varies systematically with galaxy mass in a very dramatic fashion. Such a drastic variation in M/L $(\propto M^{-2})$, is absolutely inconsistent with everything we think we know about stellar populations. Moreover, any modification attached to a length

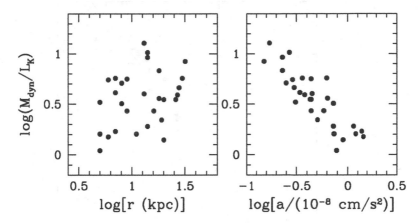

FIGURE 1. The global mass-to-K' band luminosity of Ursa Major spirals at the last measured point of the rotation curve plotted first against the radial extent of the rotation curve (left) and then against the centripetal acceleration at that point (right).

scale would imply that larger galaxies should exhibit a larger discrepancy. Anyone who has considered galaxy rotation curves knows that this is totally inconsistent with the observations. There are very small, usually low surface brightness (LSB) galaxies with large discrepancies, and very large high surface brightness (HSB) spiral galaxies with very small discrepancies.

This is shown in the first figure. At the left is a log-log plot of the dynamical $M/L_{K'}$ vs. the radius at the last measured point of the rotation curve for a uniform sample of spiral galaxies in the Ursa Major cluster (Tully et al. 1996, Verheijen and Sancisi, 2001). The dynamical M/L is calculated simply using the Newtonian formula for the mass $v^2 r/G$ (assuming a spherical mass distribution) where r is the radial extent of the rotation curve. Population synthesis studies suggest that $M/L_{K'}$ should be about one, so anything much above one indicates a discrepancy—a dark matter problem. It is evident that there is not much of a correlation of M/L with size. On the other hand, the Newtonian M/L plotted against centripetal acceleration (v^2/r) at the last measured point (right figure) looks rather different. There does appear to be a correlation in the sense that $M/L \propto 1/a$ for $a < 10^{-8}$ cm s^2. Any modification of gravity attached to a length scale cannot explain such observations.

2. Basics of MOND

Milgrom's insightful deduction was that the only viable sort of modification is one in which a deviation from Newton's law appears at low acceleration. (it should be recalled that data such as that shown in Figure 1 did not exist at the time of Milgrom's initial papers). Viewed as a modification of gravity, his suggestion was that the actual gravitational acceleration $\mathbf{g}$ is related to the Newtonian gravitational acceleration $\mathbf{g_n}$ as

$$\mathbf{g}\mu(|g|/a_o) = \mathbf{g_n} \quad , \tag{2.1}$$

where a_o is a new physical parameter with units of acceleration and $\mu(x)$ is a function which is unspecified but must have the asymptotic form $\mu(x) = x$ when $x \ll 1$ and $\mu(x) = 1$ where $x \gg 1$.

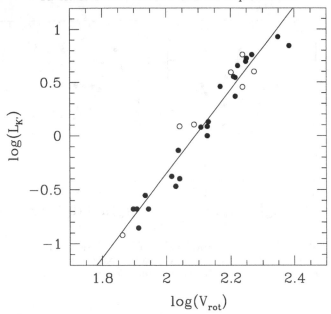

FIGURE 2. The near-infrared Tully-Fisher relation of Ursa Major spirals (Sanders & Verheijen 1998). The rotation velocity is the asymptotically constant value. The line is a least-square fit to the data and has a slope of 3.9 ± 0.2

The immediate consequence of this is that, in the limit of low accelerations, $g = \sqrt{g_n a_o}$. For a point mass M, if we set g equal to the centripetal acceleration v^2/r, this gives

$$v^4 = GMa_o \quad , \tag{2.2}$$

in the low acceleration regime. So all rotation curves are asymptotically flat and there is a mass-luminosity relation of the form $M \propto v^4$. These are aspects that are built into MOND so they cannot rightly be called predictions. However, in the context of MOND, the aspect of an asymptotically flat rotation curve is absolute. MOND leaves rather little room for maneuver; the idea is in principle falsifiable, or at least it is far more fragile than the dark matter hypothesis. Unambiguous examples of rotation curves (of isolated galaxies) which decline in a Keplerian fashion at a large distance from the visible object would falsify the idea. In effect, a rotational velocity which is constant with radius is Kepler's law in the limit of low accelerations.

In addition, the mass-rotation velocity relation and implied Tully-Fisher relation is absolute. The TF relation should be the same for different classes of galaxies and the logarithmic slope (at least of the MASS-velocity relation) must be 4—not 3.8 or 4.2— but 4.0. Moreover, it must be the case that the relation is essentially one between the total baryonic mass of a galaxy and the asymptotic flat rotational velocity—not the peak rotation velocity but the velocity at large distance. This is the most immediate and most obvious prediction (see McGaugh & de Blok 1998b and McGaugh et al. 2000 for a discussion of these points).

Converting the $M - V$ relation to the observed luminosity-velocity relation we find

$$\log(L) = 4\log(V) - \log(Ga_o < M/L >) \quad . \tag{2.3}$$

The near-infrared TF relation for Verheijen's UMa sample is shown in Figure 2 (Sanders & Verheijen 1998) where the velocity is that of the flat part of the rotation curve. The scatter about the least-square fit line of slope 3.9 ± 0.2 is consistent with observational uncertainties (i.e. no intrinsic scatter). Given the mean M/L in a particular band (≈ 1 in the K$'$ band), this observed TF relation (eq. 2.3) tells us that a_o must be on the order of 10^{-8} cm s^2. It was immediately noticed by Milgrom that $a_o \approx cH_o$ to within a factor of 5 or 6. This cosmic coincidence is quite interesting and suggests that MOND, if it is right, may reflect the effect of cosmology on local particle dynamics.

3. Implications

There are several other immediate consequences of modified dynamics—all of which were explored by Milgrom in his original papers—which do fall in the category of predictions.

(a) There exist a critical value of the surface density

$$\Sigma_c \approx a_o/G \quad . \tag{3.1}$$

If a system, such as a spiral galaxy has a surface density of matter greater than Σ_c, that means that the internal accelerations are greater than a_o, so the system is in the Newtonian regime. In systems with $\Sigma \geq \Sigma_c$ (HSB galaxies) there should be a small discrepancy between the visible and classical Newtonian dynamical mass within the optical disk. In the parlance of rotation curve observers, a HSB galaxy should be well-represented by the "maximum disk" solution (Sancisi, this volume). But in LSB galaxies ($\Sigma \ll \Sigma_c$) there is a low internal acceleration, so the discrepancy between the visible and dynamical mass would be large. These objects should be far from maximum disk. In effect, Milgrom predicted, before the actual discovery of LSB galaxies, that there would be a serious discrepancy between the observable and dynamical mass within the luminous disk of such systems—should they exist. They do exist, and this prediction has been verified—as is evident from the work of McGaugh & de Blok (1998a,b).

(b) It is well-known since the work of Ostriker & Peebles (1973), that rotationally supported Newtonian systems tend to be unstable to global non-axisymmetric modes which lead to bar formation and rapid heating of the system. In the context of MOND, these systems would be those with $\Sigma > \Sigma_c$, so this would suggest that Σ_c should appear as an upper limit on the surface density of rotationally supported systems. This critical surface density is 0.2 g cm^2 or 860 M$_\odot$/pc^2. A more appropriate value of the mean surface density within an effective radius would be $\Sigma_c/2\pi$ or 140 M$_\odot$/pc^2, and, taking $M/L_b \approx 2$, this would correspond to a surface brightness of about 22 mag/arc sec^2. There is such an observed upper limit on the mean surface brightness of spiral galaxies and this is known as Freeman's law (Freeman 1970, Allen & Shu 1979). The point is that the exitance of such a preferred surface density becomes understandable in the context of MOND.

(c) Spiral galaxies with a mean surface density near this limit—HSB galaxies—would be, within the optical disk, in the Newtonian regime. So one would expect that the rotation curve would decline in a near Keplerian fashion to the asymptotic constant value. In LSB galaxies, with mean surface density below Σ_c, the prediction is that rotation curves would rise to the final asymptotic flat value. So there should be a general difference in rotation curve shapes between LSB and HSB galaxies. In Figure 3 I show the rotation curves of two galaxies, a LSB and HSB, where we see exactly this trend. This general effect in observed rotation curves was first noted by Casertano & van Gorkom (1991).

(d) With Newtonian dynamics, pressure-supported systems which are nearly isothermal have infinite extent. But in the context of MOND it is straightforward to demonstrate that such isothermal systems are finite with the density at large radii falling roughly like $1/r^4$ (Milgrom 1984). The equation of hydrostatic equilibrium for an isotropic, isothermal

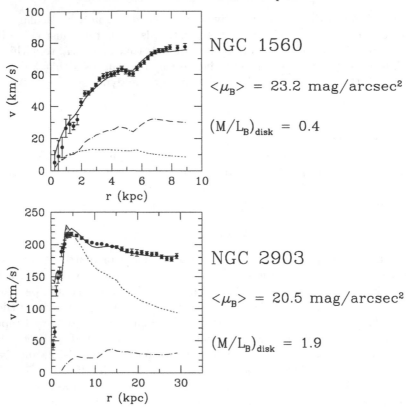

FIGURE 3. The points show the observed 21 cm line rotation curves of a low surface brightness galaxy, NGC 1560 (Broeils 1992) and a high surface brightness galaxy, NGC 2903 (Begeman 1987). The dotted and dashed lines are the Newtonian rotation curves of the visible and gaseous components of the disk and the solid line is the MOND rotation curve with $a_o = 1.2 \times 10^{-8}$ cm s^2—the value derived from the rotation curves of 10 nearby galaxies (Begeman et al. 1991). Here the only free parameter is the mass-to-light ratio of the visible component.

system reads

$$\sigma_r^2 \frac{d\rho}{dr} = -\rho g \quad , \tag{3.2}$$

where, in the limit of low accelerations $g = \sqrt{GMa_o}/r$. Here σ_r is the radial velocity dispersion and ρ is the mass density. It then follows immediately that, in this MOND limit,

$$\sigma_r^4 = GMa_o \left(\frac{d\,ln(\rho)}{d\,ln(r)}\right)^{-2} \quad . \tag{3.3}$$

Thus there exists a mass-velocity dispersion relation of the form

$$(M/10^{11}\ \mathrm{M_\odot}) \approx \left(\sigma_r/100\ \mathrm{km\ s^{-1}}\right)^4 \quad ,$$

which is similar to the observed Faber-Jackson relation (luminosity-velocity dispersion relation) for elliptical galaxies (Faber & Jackson 1976). This means that a MOND near-isothermal sphere with a velocity dispersion of 100 km s^{-1} to 300 km s^{-1} will always have a galactic mass. This is not true of Newtonian pressure-supported objects. Because of the appearance of an additional dimensional constant, a_o, in the structure equation (eq. 3.2), MOND systems are much more constrained than their Newtonian counterparts.

But with respect to actual pressure supported systems, an even stronger statement can be made. Any isolated system which is nearly isothermal will be a MOND object. That is because a Newtonian isothermal system (with large internal accelerations) is an object of infinite size and will always extend to the region of low accelerations ($< a_o$). At that point ($r_e^2 \approx GM/a_o$), MOND intervenes and the system will be truncated. This means that the internal acceleration of any isolated isothermal system (σ_r^2/r_e) is expected to be on the order of or less than a_o and that the mean surface density within r_e will typically be Σ_c or less (there are low-density solutions for MOND isothermal spheres, $\rho \ll a_o^2/G\sigma^2$, with internal accelerations less than a_o). It has been known for some time that elliptical galaxies do have a characteristic surface brightness (Fish 1964). But the above arguments imply that the same should be true of any pressure supported, near-isothermal system, from globular clusters to clusters of galaxies. Moreover, the same $M - \sigma$ relation (eq. 3.3) should apply to all such systems, albeit with considerable scatter due to deviations from a strictly isotropic, isothermal velocity field (Sanders 2000). Such deviations will also result in a dispersion of mean internal accelerations about the fiducial value of a_o.

4. Rotation curve analysis

Perhaps the most remarkable phenomenological success of MOND is in predicting the form of rotation curves from the observed distribution of detectable matter—stars and gas (Begeman et al. 1991, McGaugh & de Blok 1998, Sanders & Verheijen 1998). The procedure followed can be outlined as follows:

(a) One assumes that light traces mass, i.e. $M/L =$ constant. There are color gradients in spiral galaxies so this cannot be generally true—or at least one must decide which color band is the best tracer of the mass distribution. The general opinion is that the near-infrared emission of spiral galaxies is the optimal tracer of the underlying stellar mass distribution, since the old population of low mass stars contribute to this emission and the near-infrared is less affected by dust obscuration. So where available, near infrared surface photometry is to be preferred.

(b) In determining the distribution of detectable matter one must include the observed neutral hydrogen with an appropriate correction for the contribution of primordial helium. The gas can make a dominant contribution to the total mass surface density in some (generally low luminosity) galaxies.

(c) Given the observed distribution of mass, g_n, the Newtonian gravitational force, is calculated via the classical Poisson equation. Here it is usually assumed that the stellar and gaseous disks are razor thin. It may also be necessary to add a spheroidal bulge if the light distribution indicates the presence of such a component.

(d) Given the radial distribution of the Newtonian force, the true gravitational force, g, is calculated from the MOND formula with a_o fixed. Then the mass of the stellar disk is adjusted until the best fit to the observed rotation curve is achieved. This gives M/L of the disk as the single free parameter of the fit (unless a bulge is present).

In comparing to the observed rotation curve one assumes that the motion of the gas is co-planer rotation about the center of the given galaxy. This is certainly not always the case because there are well-known distortions to the velocity field in spiral galaxies caused by bars and warping of the gas layer. In a fully 2-dimensional velocity field these distortions can often be modeled, but the optimal rotation curves are those in which there is no evidence for the presence of significant deviations from co-planer circular motion. In general it should be remembered that not all observed rotation curves are perfect tracers

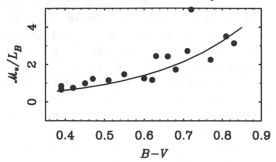
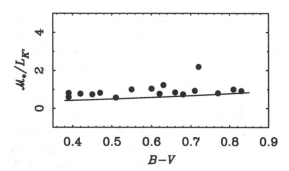

FIGURE 4. MOND inferred mass-to-light ratios for the UMa spirals (Sanders & Verheijen) in the B-band (top) and the K'-band (bottom) plotted against $B - V$ colors (McGaugh, private communication). The solid lines show predictions from populations synthesis models by Bell and de Jong (2001).

of the radial distribution of force. A perfect theory will not fit all rotation curves because of these possible problems (the same is true of a specified dark matter halo). The point is that with MOND, usually, there is one free parameter per galaxy and that is the mass or M/L of the stellar disk.

I am only going to show two examples of MOND fits to rotation curves, and these are the two galaxies already shown in Figure 3. The dotted and dashed curves are the Newtonian rotation curves of the stellar and gaseous disks respectively, and the solid curve is the MOND rotation curve with $a_o = 1.2 \times 10^{-8}$ cm s^2. We see that, not only does MOND predict the general trend for LSB and HSB galaxies, but it also predicts the observed rotation curves *in detail* from the observed distribution of matter. This procedure has been carried out for about 100 rotation curves and in only about 10 cases is the predicted rotation curve significantly different from the observed curve. For these objects there is usually an obvious problem with the observed curve or its use as a tracer of the radial force distribution.

I have noted that the only free parameter in these fits is the mass-to-light ratio of the visible disk, so one may well ask if the inferred values are reasonable. Here it is useful to consider again the Verheijen UMa sample because all galaxies are at the same distance and there is K$'$-band (near infrared) surface photometry of the entire sample. The sample also contains both HSB and LSB galaxies. Figure 5 shows the M/L in the B-band required by the MOND fits plotted against $B - V$ color (top) and the same for the K$'$-band (bottom). We see that in the K$'$-band $M/L \approx 1$ with a 30% scatter. In other words, if one were to assume a K$'$-band M/L of one at the outset, most rotation curves

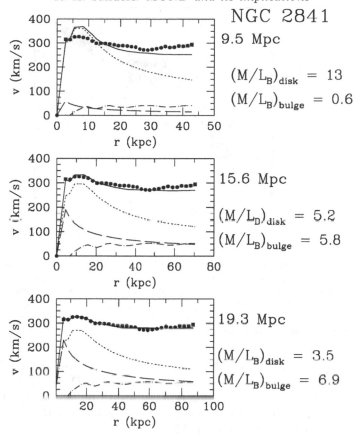

NGC 2841

9.5 Mpc

$(M/L_B)_{disk} = 13$

$(M/L_B)_{bulge} = 0.6$

15.6 Mpc

$(M/L_D)_{disk} = 5.2$

$(M/L_B)_{bulge} = 5.8$

19.3 Mpc

$(M/L_B)_{disk} = 3.5$

$(M/L_B)_{bulge} = 6.9$

FIGURE 5. MOND fits to NGC 2841 at various distances. The Hubble law distance is 9.3 Mpc ($h = 0.75$), but MOND prefers a distance of 19.3 Mpc. The Cepheid distance is 14.1 ± 1.5. The MOND rotation curve at the Cepheid distance $+1\sigma$ (15.6 Mpc) is acceptable, particularly considering the complication of the large warp in the outer regions.

would be quite precisely predicted from the observed light and gas distribution with no free parameters. In the B-band, on the other hand, the MOND M/L does appear to be a function of color in the sense that redder objects have larger M/L values. This is exactly what is expected from population synthesis models as is shown by the solid lines in both panels (Bell & de Jong 2000). This is quite interesting because there is nothing built into MOND which would require that redder galaxies should have a higher M/L_b; this simply follows from the rotation curve fits.

We sometimes hear that it is not so surprising that MOND fits rotation curves because that is what it was designed to do. This is certainly not correct. MOND was designed to produce asymptotically flat rotation curves with a given mass-velocity relation (or TF law). It was not designed to fit the details of all rotation curves with a single adjustable parameter (even of galaxies which are gas-dominated with no adjustable parameter), and it was certainly not designed to provide a reasonable dependence of fitted M/L on color. Indeed, there are a couple of well-observed spiral galaxies which are problematic for MOND, and which could, in principle, falsify the idea. One of these is NGC 2841—a large spiral galaxy with a Hubble distance of about 9 Mpc (Begeman et al. 1991). In fact, the rotation curve of the galaxy cannot be fit using MOND if the distance is only 9 Mpc. MOND prefers a distance of 19 Mpc as we see in Figure 5 (the scaling of the

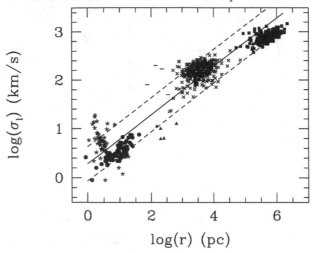

FIGURE 6. The line-of-sight velocity dispersion vs. characteristic radius for pressure-supported astronomical systems. The star-shaped points are globular clusters (Trager et al. 1993), the points are massive molecular clouds in the Galaxy (Solomon et al. 1987), the crosses are massive elliptical galaxies (Jorgensen et al. 1995a,b), and the squares are X-ray emitting clusters of galaxies (White, Jones & Forman 1997). The solid line is shows the relation $\sigma_l^2/r = a_o$ and the dashed lines a factor of 5 variation about this relation.

centripetal acceleration depends upon the distance). If the distance to this galaxy is really less than about 14 Mpc it is quite problematic for MOND. Now it turns out that a Cepheid distance to this galaxy has just been determined (Macri et al. 2001), and this is 14.1 ± 1.5 Mpc. Given that the distance could easily be as large as 15.6 Mpc, this galaxy now would seem to present no problem for MOND.

The success of MOND in accounting for galaxy rotation curves with only one free parameter, the M/L of the visible disk which usually assumes quite reasonable values, is remarkable. *Whether MOND is correct or not, the success of this simple algorithm implies that galaxy rotation curves are entirely determined by the distribution of visible matter. If you believe in dark matter, then you somehow must explain this phenomenology. How can the distribution of dark matter be so intimately connected with the distribution of visible matter?*

5. Pressure-supported systems

Figure 6 is a log-log plot of the velocity dispersion versus size for pressure-supported, nearly isothermal astronomical systems. At the bottom of the plot are globular clusters (star-shaped) and giant molecular clouds (points) in the Galaxy. The group of points in the middle are ellipticals (crosses) and at the top are X-ray emitting clusters of galaxies (squares). The triangles are the dwarf spheroidal systems surrounding the Milky Way and the dashes are compact dwarf ellipticals. The plotted parameters have not been massaged at all but are taken directly from the relevant observational papers. The measure of size is not homogeneous—for ellipticals and globular clusters it is the well-known effective radius, for the X-ray clusters it is an X-ray intensity isophotal radius, and for the molecular clouds it is a isophotal radius of CO emission. The velocity dispersion refers to the central velocity dispersion for ellipticals and globulars; for the clusters it is the thermal velocity dispersion of the hot gas; for the molecular clouds it is just the typical line width of the CO emission. A velocity-dispersion–size correlation has been previously claimed for individual classes of objects—most notably, the molecular clouds and the clusters of galaxies.

The parallel lines are not fits but represent fixed internal accelerations. The solid line corresponds to $\sigma_l^2/r = 10^{-8}$ cm s^2 and the parallel dashed lines to accelerations 5 times larger or smaller than this particular value. It is clear from this diagram that the internal accelerations in these systems all lie within a factor of a few of a_o. This also implies that the surface densities in these systems are near the MOND surface density Σ_c.

So these astronomical objects appear to have a characteristic internal acceleration or a characteristic surface density as MOND predicts. I emphasize that these objects are not only pressure-supported, but they are also nearly isothermal; i.e. there is not a large variation in the line-of-sight velocity dispersion across these objects. Stars are also pressure-supported systems but they would lie far outside the upper left boundary of this plot. However, stars are very far from isothermal.

It has been noted above that, with MOND, such self-gravitating near-isothermal systems would be expected to have internal accelerations comparable to or less than a_o. *But it is not at all evident how Newtonian theory can account for the fact these different classes of astronomical objects, covering a large range in size and located in very different environments, all appear to have comparable internal accelerations near the cosmologically interesting value of cH_o.*

With MOND, systems that lie below the line, i.e. with low internal accelerations, would be expected to exhibit larger discrepancies. This is particularly true of the dwarf spheroidal systems. Systems above the line (ellipticals) are high surface brightness systems and if interpreted in terms of Newtonian dynamics, would not exhibit much need for dark matter inside an effective radius. This seems to be the case. I just add that the MOND M-σ relation (eq. 3.3) is very sensitive to variations from strict homology which would be expected to lead to a large scatter in the observed Faber-Jackson law. However, MOND imposes boundary conditions on the inner Newtonian solution which restrict non-homologous objects to lie on a narrow fundamental plane similar to that implied by the traditional virial theorem (Sanders 2000).

Note that clusters of galaxies lie below the $\sigma_l^2/r = a_o$ line in Figure 6; thus, these objects would be expected to exhibit significant discrepancies. That this is the case has been known for 70 years (Zwicky 1933), although the subsequent discovery of hot X-ray emitting gas goes some way in alleviating the original discrepancy. For an isothermal sphere of hot gas at temperature T, the Newtonian dynamical mass within radius r_o, calculated from the equation of hydrostatic equilibrium, is

$$M_n = \frac{r_o}{G}\frac{kT}{m}\left(\frac{d\ln(\rho)}{d\ln(r)}\right) \quad , \tag{5.1}$$

where m is the mean atomic mass and the logarithmic density gradient is evaluated at r_o. For the X-ray clusters plotted in Fig. 6 this turns out to be typically about a factor of 4 or 5 larger than the observed mass in hot gas and in the stellar content of the galaxies. This rather modest discrepancy viewed in terms of dark matter has led to the so-called baryon catastrophe—not enough non-baryonic dark matter in the context of standard CDM cosmology (White et al. 1993).

With MOND, the dynamical mass (eq. 3.3) is given by

$$M_m = (Ga_o)^{-1}\left(\frac{kT}{m}\right)^2\left(\frac{d\ln(\rho)}{d\ln(r)}\right)^2 \quad , \tag{5.2}$$

and the discrepancy, using the same value of a_o determined from nearby galaxy rotation curves, is on average reduced to about a factor of 2 larger than the observed mass. There does indeed seem to be a remaining discrepancy. This could be interpreted as a failure, or one could say that MOND predicts that the baryonic mass budget of clusters is not yet complete and that there is more mass to be detected (Sanders 1999). It would have certainly been devastating for MOND had the predicted mass turned out to be typically *less* than the observed mass in hot gas and stars.

6. Cosmology and structure growth

Let me just summarize what I have said so far. MOND not only allows the form of rotation curves to be precisely predicted from the distribution of observable matter, but it also explains certain systematic aspects of the photometry and kinematics of galaxies and clusters: the presence of a preferred surface density in spiral galaxies and ellipticals— the so-called Freeman and Fish laws; the fact that pressure-supported nearly isothermal systems ranging from molecular clouds to clusters of galaxies are characterized by a specific internal acceleration (a_o); the exitance of a TF relation with small scatter— specifically a correlation between the baryonic mass and the asymptotically flat rotation velocity of the form $v^4 \propto M$; the Faber-Jackson relation for ellipticals, and with more detailed modeling, the Fundamental Plane; not only the magnitude of the discrepancy in clusters of galaxies but also the fact that mass-velocity dispersion relation which applies to elliptical galaxies (eq. 3.2) extends to clusters (the mass-temperature relation). And it accomplishes all of this with a single new parameter with units of acceleration—a parameter determined from galaxy rotation curves which is within an order of magnitude of the cosmologically significant value of cH_o. This is why several of us believe that, on an epistemological level, MOND is more successful than dark matter.

But, of course, MOND must fit into a larger picture. One may naturally ask—what are the larger-scale implications of modified dynamics—specifically what are implications for gravitational lensing and does MOND imply a reasonable cosmology and cosmogony? These are questions which require a more basic theory underlying MOND, and this is, at present, the essential weakness of the idea.

Frequently, the absence of a covariant theory is presented as an argument against MOND. But the criterion for judging a scientific hypothesis surely must be its empirical success. The absence of a successful covariant version is simply an aspect of its incompleteness. People don't reject general relativity because there is not yet a viable theory of quantum gravity. At the same time, it is fair to say that MOND will never be entirely credible to most astronomers and physicists until it makes some contact with more familiar physics.

There have been several attempts to construct a more general theory, most notably by Bekenstein (1987), and while these are very nice ideas, none of these attempts is entirely satisfactory for various reasons (Bekenstein & Sanders 1994, Sanders 1997). A different approach is to consider MOND as modified inertia (Milgrom 1994), perhaps resulting from the interaction of an accelerating particle with vacuum fields (Milgrom, 1999). Here the coincidence between a_o and cH_o plays a central role: if inertia results from influence of the vacuum on accelerated motion, then, because a cosmological constant has a non-trivial effect upon the vacuum, we might expect that it also has a non-trivial effect upon inertia. It is beyond my mission to describe these ideas in detail, but I would just like to comment upon the possible shape of a MOND cosmology.

First of all, I take it that the experimental foundations of the standard Big Bang are so well-established, that any underlying theory of MOND should not lead to a radically different cosmology, at least not in the early Universe. Then, to say that MOND is an alternative to dark matter does not mean that every baryon in the Universe must be glowing with an M/L of one. In fact this is certainly not the case since Ω in visible matter is substantially less than Ω in baryons. Moreover, there are clear indications that at least some flavors of neutrinos have a non-vanishing mass (J. Bahcall, this conference), so there is a contribution of non-baryonic dark matter to the total mass budget of the Universe— at least at a level comparable to the mean density of baryons in visible stars. But it would be contrary to the spirit of MOND if dark matter—baryonic or non-baryonic—were a dominant constituent of the Universe or of bound gravitational systems such as galaxies or clusters of galaxies. So the question arises—is cosmology with $\Omega_m \approx \Omega_b \approx 0.02/h^2$ compatible with observations—in particular, with the recent Boomerang and Maxima observations of the CMB fluctuations?

This is a question that has been considered by McGaugh (1999, 2000), who applied the widely-used CMBFAST program (Seljak & Zeldarriaga 1996) in the case of a pure baryonic Universe. Before the Boomerang and Maxima results appeared (Hanany et al. 2000, Lang et al. 2001), McGaugh pointed out that a pure baryonic universe, with the dominant constituent of the Universe being in vacuum energy density, would imply that the second peak in the angular power spectrum should be much reduced with respect to the expectations of the concordance ΛCDM cosmology. The reason for this suppression is basically Silk damping (Silk 1968) in a low Ω_m, pure baryonic universe—the shorter wavelength fluctuations are exponentially suppressed by photon diffusion. When the Boomerang results appeared, much of the excitement was generated by the unexpected low amplitude of the second peak. With $\Omega_{\text{total}} = 1.01$ and $\Omega_m = \Omega_b$ (no CDM or non-baryonic matter of any sort) McGaugh produced a rather nice match to the Boomerang results. A further prediction is that the third peak should be even more reduced. There are indications from the recent more complete analyses of BOOMERANG data (Netterfield et al. 2001) that this may not be the case, but the systematic uncertainties remain large. In addition, the SNIa results on the accelerated expansion of the Universe (Perlmutter et al. 1999) as well as the statistics of gravitational lensing (Falco et al. 1998) seem to exclude a pure baryonic and vacuum energy Universe, although it is unclear that all systematic effects are well-understood. It is also possible that a MOND cosmology may differ from standard Friedmann cosmology in the low-z Universe (note that in some brane-world scenarios late-time cosmology diverges from Friedmann cosmology, e.g. Deffayet 2001).

Of course, if we live in a Universe of only baryons, then how does structure form? After all a primary motivation for non-baryonic cosmic dark matter is the necessity of forming the observed structure in the Universe by the present epoch via gravitational growth of very small density fluctuations. As we all know, non-baryonic dark matter helps because it offers the possibility that fluctuations can begin growing before the epoch of hydrogen recombination. The expectation is that MOND, by providing stronger effective gravity in the limit of low accelerations, might also help.

In the absence of a proper theory, this question can be considered by making several *Ansätze* in the spirit of the existing bits of the theory:

(*a*) The MOND acceleration parameter a_o is constant with cosmic time. This could be the case if a_o is related to the cosmological term ($a_o \approx c\sqrt{\Lambda}$).

(*b*) MOND is applied in determining the peculiar accelerations—those accelerations which develop around density perturbations—and not to the overall Hubble flow. That is to say, the Hubble flow remains intact. One might imagine that MOND should be applied to the Hubble flow; that is, as soon as the deceleration of the Hubble flow over a finite size region falls below a_o, then the dynamics of that region begins to deviate from the standard Friedmann solutions. This would lead to the eventual collapse of any finite size region regardless of its initial density and expansion velocity (Felten 1984, Sanders 1998). With this sort of cosmology the evolution of the early Universe would be as it is

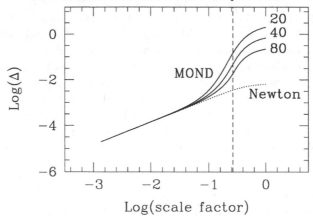

FIGURE 7. The growth of fluctuations with an initial amplitude of 10^{-5}. The solid lines show the growth of fluctuations on various comoving scales in the context of the simple non-relativistic MOND theory (Sanders 2001) and dotted line is the usual Newtonian growth in the pure baryonic Universe. The vertical dashed line indicates the scale factor at which the cosmological term begins to dominate the expansion in this model universe.

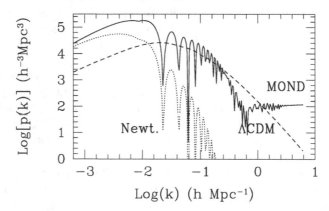

FIGURE 8. The solid line shows the power spectrum resulting from the toy MOND theory applied to the growth of fluctuations with an initial Harrison-Zeldovitch, COBE-normalized power spectrum. The dashed line is the usual ΛCDM power spectrum and the dotted line shows the power spectrum that would result from the Newtonian growth of fluctuations in the pure-baryonic Universe.

in the standard Big Bang (the deceleration on relevant scales is much larger than a_o) but the present Universe would look rather different than it actually does.

(*c*) Although the MOND does not affect the Hubble flow, the deceleration or acceleration of the Hubble flow enters as a background field which influences the development of peculiar accelerations in the MOND regime.

It is possible to construct a non-relativistic Lagrangian-based theory which incorporates these three assumptions (Sanders 2001)—this is similar to the 2-field version of the theory of Bekenstein and Milgrom (1984). Following the same procedure as in Newtonian cosmology, I find a growth equation for small density fluctuations which is non-linear even in the regime where the density fluctuations are small (this is because MOND is fundamentally non-linear). The growth of fluctuations becomes dramatically rapid when the non-linear term dominates as is evident in Figure 7 which is a plot of the fluctuation amplitude as a function of scale factor in a baryonic-vacuum energy dominated Universe. Fluctuations of smaller wavelength grow to larger amplitude because they enter the MOND regime earlier.

The non-linear term becomes important when the background acceleration vanishes, i.e. when the density in vacuum energy becomes comparable to the matter energy density. Thus in a MOND Universe we might expect structure and massive galaxies to form about when the cosmological constant begins to dominate the expansion. Starting with an initial Harrison-Zeldovitch power spectrum normalized by COBE, the final power spectrum is shown in Figure 8 where it is compared to the ΛCDM power spectrum. We see that it is quite similar apart from the baryonic oscillations.

So MOND offers the possibility of overcoming the slow growth of fluctuations in a pure baryonic Universe. It also offers an explanation of why we are observing the Universe at an epoch when Λ has only recently emerged as the dominant term in the Friedmann equation. If this scenario remains as an aspect of a fully covariant theory, then the cosmological argument is no longer a unique rationale for non-baryonic dark matter.

I am very grateful to Stacy McGaugh for useful discussions and for preparing Figure 5. I thank Mario Livio for his kind invitation to speak at this conference and for his efforts in organizing such an excellent and well-balanced scientific program.

REFERENCES

ALLEN, R. J. & SHU, F. H. 1979 *ApJ* **227**, 67.

BEGEMAN, K. G. 1989 *A&A* **223**, 47.

BEGEMAN, K. G., BROEILS, A. H., & SANDERS, R. H. 1991 *MNRAS* **249**, 523.

DOLL, E. F. & DE JONG, R. S. 2001 *ApJ* **550**, 212.

BEKENSTEIN, J. D. 1988. In *Second Canadian Conf. on General Relativity and Relativistic Astrophysics* (eds. A. Coley, C. Dyer, & T. Tupper). p. 487. World Scientific.

BEKENSTEIN, J. D. & MILGROM, M. 1984 *ApJ* **286**, 7 (BM).

BEKENSTEIN, J. D. & SANDERS, R. H. 1994 *ApJ* **286**, 7.

BROEILS, A. H. 1992 *PhD Dissertation*, Univ. of Groningen.

CASERTANO, S. & VAN GORKOM, J. H. 1991 *AJ* **101**, 1231.

DEFFAYET, C. 2001 *Phys. Lett. B* **502**, 199.

FABER, S. M. & JACKSON, R. E. 1976 *ApJ* **204**, 668.

FALCO, E. E., KOCHANEK, C. S., MUNOZ, J. A. 1998 *ApJ* **494**, 47.

FELTEN, J. E. 1984 *ApJ* **286**, 38.

FISH, R. A. 1964 *ApJ*, **139**, 284.

FREEMAN, K. C. 1970 *ApJ* **160**, 811.

HANANY, S., ET AL. 2000 *ApJ* **545**, L5.

JØRGENSEN, I., FRANX, M. & KÆRGARD, P. 1995a *MNRAS* **273**, 1097.

JØRGENSEN, I., FRANX, M. & KÆRGARD, P. 1995b *MNRAS* **276**, 1341.

LANGE, A. E., ET AL. 2001 *Phys. Rev D* **63**, 042001.

MACRI, L. M., ET AL. 2001 *astro-ph/0105491*.

McGAUGH, S. S. & DE BLOK, W. J. G. 1998a *ApJ* **499**, 66.

McGAUGH, S. S. & DE BLOK, W. J. G. 1998b *ApJ* **508**, 132.

McGAUGH, S. S. 1999 *ApJ* **523**, L99.

McGAUGH, S. S. 2000 *ApJ* **541**, L33.

McGAUGH, S. S., SCHOMBERT, J. M., BOTHUN, G. D., & DE BLOK, W. J. G. 2000 *ApJ* **533**, 99.

MILGROM, M. 1983a *ApJ* **270**, 365.

MILGROM, M. 1983b *ApJ* **270**, 371.

MILGROM, M. 1983c *ApJ* **270**, 384.

MILGROM, M. 1984 *ApJ*, **287**, 571.

MILGROM, M. 1994 *Ann. Phys.* **229**, 384.

MILGROM, M. 1999 *Phys. Lett. A* **253**, 273.

NETTERFIELD, C. B., ET AL. 2001 *astro-ph/104460*

OSTRIKER, J. P. & PEEBLES, P. J. E. 1973 *ApJ* **186**, 467.

PERLMUTTER, S., ET AL. 1999 *ApJ* **517**, 565.

SANDERS, R. H. 1997 *ApJ* **480**, 492.

SANDERS, R. H. 1998 *MNRAS* **296**, 1009.

SANDERS, R. H. 1999 *ApJ* **512**, L23.

SANDERS, R. H. 2000 *MNRAS* **313**, 767.

SANDERS, R. H. 2001 *ApJ,* **560**, 1; *astro-ph/0011439*.

SANDERS, R. H. & VERHEIJEN, M. A. W. 1998 *ApJ* **503**, 97.

SELJAK, U. & ZELDARRIAGA, M. 1996 *ApJ* **469**, 437.

SILK, J. 1968 *ApJ* **151**, 469.

SOLOMON, P. M., RIVOLO, A. R., BARRETT, J., & YAHIL, A. 1987 *ApJ* **319**, 730.

TRAGER, S. C., DJORGOVSKI, S., & KING, I. R. 1993. In *Structure and Dynamics of Globular Clusters*, APS Series vol. 50, p. 347. ASP.

TULLY, R. B. & FISHER, J. R. 1977 *A&A* **54**, 661.

TULLY, R. B., VERHEIJEN, M.A.W., PIERCE, M.J., HANG, J-S., WAINSCOT, R. 1996 *AJ* **112**, 2471.

VERHEIJEN, M. A. W. & SANCISI, R. 2001 *A&A* **370**, 765.

WHITE, S. D. M., NAVARRO, J. F., EVRARD, A. E., & FRENK, C. S. 1993 *Nature* **336**, 429.

WHITE, D. A., JONES, C., & FORMAN, W. 1997 *MNRAS* **292**, 419.

ZWICKY, F. 1933 *Helv. Phys. Acta* **6**, 110.

Cosmological parameters and quintessence from radio galaxies

By RUTH A. DALY[1]† AND ERICK J. GUERRA[2]

[1]Department of Physics, Berks-Lehigh Valley College, Penn State University, P.O. Box 7009, Reading, PA 19610-6009, USA; rdaly@psu.edu

[2]Department of Chemistry & Physics, Rowan University, Glassboro, NJ 08028-1701, USA; guerra@scherzo.rowan.edu

FRIIb radio galaxies provide a tool to determine the coordinate distance to sources at redshifts from zero to two. The coordinate distance depends on the present values of global cosmological parameters, quintessence, and the equation of state of quintessence. The coordinate distance provides one of the cleanest determinations of global cosmological parameters because it does not depend on the clustering properties of any of the mass-energy components present in the universe.

Two complementary methods that provide direct determinations of the coordinate distance to sources with redshifts out to one or two are the modified standard yardstick method utilizing FRIIb radio galaxies, and the modified standard candle method utilizing type Ia supernovae. These two methods are compared here, and are found to be complementary in many ways. The two methods do differ in some regards; perhaps the most significant difference is that the radio galaxy method is completely independent of the local distance scale and independent of the properties of local sources, while the supernovae method is very closely tied to the local distance scale and the properties of local sources.

FRIIb radio galaxies provide one of the very few reliable probes of the coordinate distance to sources with redshifts out to two. This method indicates that the current value of the density parameter in non-relativistic matter, Ω_m, must be low, irrespective of whether the universe is spatially flat, and of whether a significant cosmological constant or quintessence pervades the universe at the present epoch.

The effect of quintessence, with equation of state w, is considered. FRIIb radio galaxies indicate that the universe is currently accelerating in its expansion if the primary components of the universe at the present epoch are non-relativistic matter and quintessence, and the universe is spatially flat.

---

1. Introduction

Current values of global cosmological parameters that control and describe the state and expansion rate of the universe are still not known with certainty. The components of the universe at the present epoch can be put into three categories: non-relativistic matter; photons and neutrinos (important in the early universe); and a third component that has yet to be identified, and may be a cosmological constant, quintessence, space curvature, or something else.

Non-relativistic matter includes baryons and the dark matter known to cluster with galaxies and galaxy clusters; the total, normalized, mass density contributed by this component at present is Ω_m. Non-relativistic matter is known to play an important role in the expansion rate of the universe at the present epoch. Photons that make up the cosmic microwave background and neutrinos produced in the early universe are known with some certainty, but do not contribute significantly to the mass-energy density at present, and hence do not play an important role in controlling the expansion rate of the universe at the present epoch.

† NSF National Young Investigator

There are numerous indications that Ω_m is low; in fact, radio galaxies alone indicate that Ω_m must be less than about 0.63 at about 95% confidence, and radio galaxies alone indicate that Ω_m equal to unity is ruled out at about 99% confidence (Daly & Guerra 2002 [DG02]; Guerra, Daly, & Wan 2000 [GDW00]). This includes possible contributes from space curvature, a cosmological constant, or quintessence. Many methods indicate that Ω_m is low (e.g. Turner & White 1997; Perlmutter, Turner, & White 1999; Wang et al. 2000).

If $\Omega_m < 1$, then there must be a third component, and this component must play a significant role in determining the expansion rate of the universe at the present epoch. This component could be space curvature, or could be a component of mass-energy with an equation of state $w = P/\rho$ that is different from non-relativistic matter, which has $w = 0$; here P is the pressure of the component, and ρ its mass-energy density. Hopefully, there is only one unknown component that is significant at the present epoch.

Recent observations of fluctuations of the microwave background radiation at the last scattering surface indicate that space curvature is close to zero (de Bernardis et al. 2000, Balbi et al. 2000, Bond et al. 2000). This simplifies the determination of the third component. A general form for the third component, referred to as quintessence, allows constraints on both the normalized mass-energy density $\Omega_Q = 1 - \Omega_m$, and equation of state w. Constraints imposed by FRIIb radio galaxies on quintessence are presented in Section 4. FRIIb radio galaxies are the most powerful FRII sources; they have regular radio bridge structure indicating an average growth rate that is supersonic (see Daly 2001; GDW00, and Wan, Daly, & Guerra 2000 [WDG00]).

The outline of the paper is as follows. In Section 2, the radio galaxy and supernova methods are compared. In Section 3, the radio galaxy method is described more fully. The method is applied to a spatially flat universe with quintessence in Section 4; it is shown that the radio galaxy method indicates that the expansion of the universe is accelerating at present. In addition, the radio galaxy method is applied to a universe that may have space curvature, a cosmological constant, and non-relativistic matter; it is shown that radio galaxies indicate that Ω_m must be low.

The application of radio galaxies as a modified standard yardstick not only yields constraints on global cosmological parameters and quintessence, but also yields constraints on models of energy extraction from massive black holes, as discussed in Section 5. FRIIb radio sources also allow a determination of the pressure, density, and temperature of the gas around the source, as discussed in Section 6. Studies of FRIIb sources indicate that they are in the cores of clusters or proto-clusters of galaxies, and thus provide a way to study evolution of structure, and a separate way to constrain cosmological parameters. Conclusions are presented in Section 7.

2. Comparison of radio galaxy and supernova methods

Constraints on global cosmological parameters through the determination of the co-ordinate distance to high-redshift sources is particularly clean since it depends only on global cosmological parameters such as Ω_m and Ω_Λ allowing for space curvature, or Ω_m and the equation of state of quintessence w assuming zero space curvature and allowing for quintessence. The coordinate distance does *not* depend on how the mass is clustered, the properties of the initial fluctuations, how density fluctuations evolve, whether baryonic and clustered dark matter are biased, and a whole host of issues that confound other methods of determining global cosmological parameters. The only assumption is that the different mass-energy components are homogeneous and isotropic on large scales, scales much smaller than the scale of the coordinate distance being determined.

Supernovae	Radio Galaxies
Type SNIa	Type FRIIb
$\propto (a_o r)^{2.0}$	$\propto (a_o r)^{1.6}$
$0 < z < 1$	$0 < z < 2$
~ 100 sources	20 sources (70 in parent pop.)
modified standard candle	modified standard yardstick
light curve $\Longrightarrow$ peak luminosity	radio bridge $\Longrightarrow$ average length
empirical relation	physical relation
normalized at $z = 0$	not normalized at $z = 0$
(depends on local distance scale)	(independent of local distance scale)
depends on local source properties	independent of local sources
	Ω_m is low
universe is accelerating	universe is accelerating ($\sim 2\ \sigma$)
some theoretical understanding	good theoretical understanding
well-tested empirically	needs more empirical testing

TABLE 1. Comparison of supernova and radio galaxy methods

Two methods currently being used to determine the coordinate distance to high-redshift sources are the radio galaxy method and the supernova method. Table 1 lists a comparison of key aspects of each method. The supernova method uses type Ia supernovae as a modified standard candle, as summarized in the papers of Riess et al. (1998) and Perlmutter et al. (1999). The rate of decline of the light curve is used to predict the peak luminosity of a supernova; the predicted peak luminosity and the observed peak flux density are used to determine $(a_o r)^{0}$, where $(a_o r)$ is the coordinate distance to the source. (Recall that the luminosity distance is $d_L = (a_o r)(1 + z)$, and the source redshifts are all known.)

The radio galaxy method uses FRIIb radio galaxies as a modified standard yardstick, and is described in more detail in Section 3. The method relies on a comparison between an individual source and the properties of the parent population at the source redshift. The radio surface brightness and width of the radio source can be used to predict the maximum source size, measured using the largest angular size of the source or the separation between the radio hotspots. The average source size D_* is just half of the maximum source size predicted using the model. This determination of the average source length does not depend on the observed length of the source, D, and depends on the coordinate distance to the $(-2\beta/3) + (4/7)$ power: $D_* \propto (a_o r)^{(-2\beta/3)+(4/7)}$. The predicted average size of a given source is equated to the average size of all FRIIb radio galaxies at similar redshift $<D>$, and, of course, $<D> \propto (a_o r)$. Clearly the two measures of the average source size must be equal, or their ratio must be a constant, independent of redshift: $<D> /D_* =$ constant. And, their ratio depends on $(a_o r)^{(2\beta/3)+(3/7)}$. Thus, this method allows a determination of the model parameter β and global cosmological parameters. It turns out that β is relevant to models of jet formation and energy extraction from AGN, as discussed in Section 5, and by Daly & Guerra (2002) [DG02].

The radio galaxy and supernova methods are complementary, as shown in Table 1. They have a similar dependence on the coordinate distance; the power listed for radio galaxies is for a value of β of 1.75, but this power does not change by very much given the range of β allowed by the data ($\beta = 1.75 \pm 0.25$). They cover similar redshift ranges, though the radio galaxy method does go to redshifts of two. They have a similar number of sources, though the supernova method does have more sources than the radio galaxy method, and thus is much more well tested empirically, as noted in the table. The su-

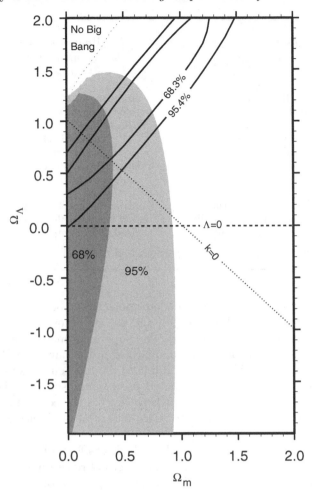

FIGURE 1. Two-dimensional constraints obtained using radio galaxies (shaded regions) compared with those from obtained by the supernovae cosmology team using supernovae (solid lines)(see Perlmutter et al. 1999 and GDW00). One-dimensional constraints obtained using radio galaxies are discussed in the text.

pernova method relies on a relation that is derived empirically, while the radio galaxy method relies on a relation that is derived using physical and theoretical arguments (see Section 3). The supernova method is normalized at zero redshift, and depends quite strongly on the local distance scale and properties of local supernova. The radio galaxy method is not normalized at zero redshift, is completely independent of the local distance scale, and is independent of the properties of local sources. GDW00 show that the analysis can be carried using sources with redshifts greater than 0.3 only, and the results are identical to those obtained including sources with redshifts less than 0.3; this is also shown here in Figures 4, 5, and 8. The radio galaxy method indicates that Ω_m must be low irrespective of the properties of the unknown component at zero redshift. This method also implies that if space curvature is zero, then it is quite likely (84% confidence) that the universe is accelerating at the present epoch (see Figure 9). The supernova method indicates that the universe is accelerating at the present epoch. The supernova method is well-tested empirically, but needs more theoretical understanding, while the radio galaxy method is better understood theoretically, and needs more empirical testing.

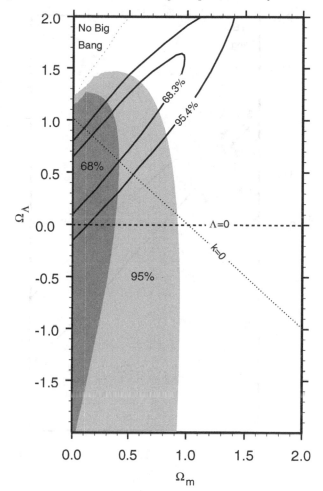

FIGURE 2. Two-dimensional constraints obtained using radio galaxies (shaded regions) compared with those obtained by the high-redshift supernovae team using supernovae (solid lines) (see Riess et al. 1998 and GDW00). One-dimensional constraints obtained using radio galaxies are discussed in the text.

The supernova method relies on relatively short-lived optical events on the scale of stars, while the radio galaxy method relies on much more long-lived radio events on scales larger than the scale of galaxies. Thus, any possible selection effects or unknown systematic errors must be completely different for the two methods. The fact that they yield such similar results suggests that both are accurate, and are not plagued by unidentified errors.

The two-dimensional results obtained using radio galaxies are compared with those obtained using supernova and the cosmic microwave background allowing for space curvature, a cosmological constant, and non-relativistic matter, and are shown in Figures 1, 2, and 3. Note, that the contours shown are a joint probability, so constraints on cosmological parameters can not be read directly from these figures. To read off constraints on individual cosmological parameters directly from a figure, the one-dimensional figure must be considered. For the radio galaxy method, these are presented in Figures 4 and 5, and in GDW00 and DG02. The radio galaxy sample includes 20 FRIIb galaxies for which D_* has been computed, and the 70 FRIIb radio galaxies in the parent population. The

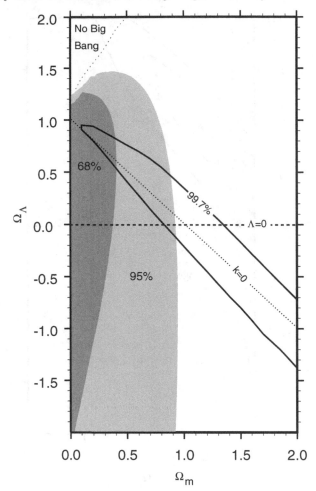

FIGURE 3. Two-dimensional constraints from radio galaxies (shaded regions) compared with those from CMB anisotropy (solid lines) (see Bond et al. 2000 and GDW00). The CMB anisotropy measurements indicate that $k = 0$.

20 sources for which D_* has been determined are a subset of the 70 sources that make up the parent population.

The fact that the methods constrain different parts of the Ω_m-Ω_Λ plane is related to the fact that the methods cover different redshift ranges (e.g. Reiss 2000). Most of the supernovae data points are at redshifts less than about one or so. The radio galaxies are primarily at redshifts from 0.5 to 2, and the cosmic microwave background is at a redshift of about a thousand. Thus, the three data sets are complementary in their redshift coverage, and the constraints they impose.

3. FRIIb radio galaxies as a modified standard yardstick

A detailed description of FRIIb radio galaxies as a tool to determine global cosmological parameters may be found in Daly (1994), Guerra (1997), Guerra & Daly (1998), GDW00, DG02.

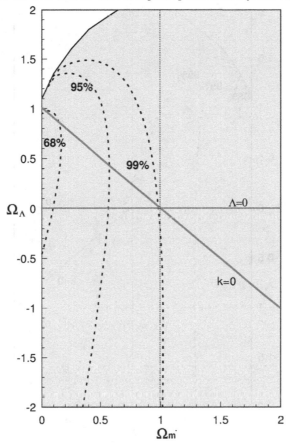

FIGURE 4. Constraint obtained using FRIIb radio galaxies. The projections of the 68%, 95%, and 99% confidence intervals onto either axis (Ω_m or Ω_Λ) indicates the probability associated with the range in that one parameter, independent of all other parameter choices. That is, these are the one-dimensional confidence contours (see GDW00).

FRIIb radio galaxies have very regular bridge structure. They are the most powerful FRII sources, having radio powers about a factor of 10 above the classical FRI–FRII separation. The regular bridge structure indicates that both the instantaneous and time-average rate of growth of the source is supersonic (e.g. Daly 2001).

The method is based on the following premises and assumptions: the forward region of FRIIb radio galaxies are governed by strong shock physics; the total source lifetime t_* is related to the beam power L_j via the relation $t_* \propto L_j^{-\beta/3}$; and the sources in the parent population at a given redshift have a similar maximum or average size. If these conditions are satisfied, then FRIIb radio galaxies provide a reliable probe of global cosmological parameters. Observations of FRIIb sources indicate that the forward region (near the radio hotspot) are governed by strong shock physics. A power-law relation between the beam power and the source lifetime is expected/predicted in currently popular models to produce large scale jets (DG02). And, the dispersion in source size at a given redshift for the parent population of sources suggests that the sources at a given redshift do have a similar average size (see Figure 6).

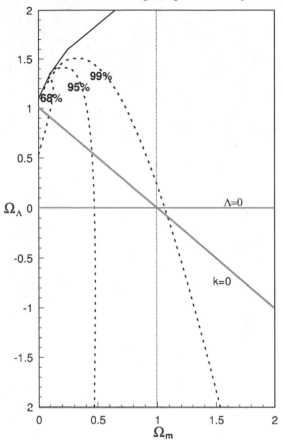

FIGURE 5. The same as Figure 4, excluding the lowest redshift bin (i.e. Cygnus A), from the fit. Only sources with redshifts greater than 0.3 are included in this analysis (see GDW00). The results are nearly identical to those presented in Figure 4.

From these simple assumptions, the average size a given source will have if it could be observed over its entire lifetime is

$$D_* \propto (P_L a_L^2)^{-\beta/3} \, v_L^{1-\beta/3} \propto (a_o r)^{-2\beta/3+4/7} \quad , \tag{3.1}$$

where each contributor to D_* can be determined from radio bridge observations including the lobe pressure P_L, the lobe width a_L, and the average rate at which the bridge is lengthening v_L.

Minimum energy conditions are not assumed to hold in the bridges of the sources; an offset from minimum energy conditions is included, and is one of the parameters that is obtained from the analysis (see Wellman, Daly, & Wan 1997b [WDW97b]).

The average size of a given source D_* is assumed to be equal to the average size of the parent population at the same redshift $<D>$, where $<D> \propto (a_o r)$. Thus, the ratio $<D>/D_*$, which must be a constant, can be used to determine the coordinate distance to the source since

$$<D>/D_* \propto (a_o r)^{2\beta/3+3/7} \quad . \tag{3.2}$$

Figure 6 shows $<D>$ as a function of redshift, and Figure 7 shows D_* as a function of redshift. The method relies on the fact that $D_* \propto (a_o r)^{-0.6}$ for $\beta = 1.75$, while $<D> \propto (a_o r)$. Thus, the method boils down to finding the cosmological parameters such

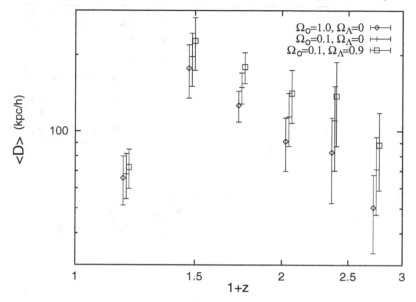

FIGURE 6. The average lobe-lobe size $<D>$ of powerful extended 3CR radio galaxies for the 70 FRIIb radio galaxies that comprise the parent population of radio galaxies. Different choices of cosmological parameters are shown (here $\Omega_m = \Omega_o$). Note that $<D> \propto (a_o r)$.

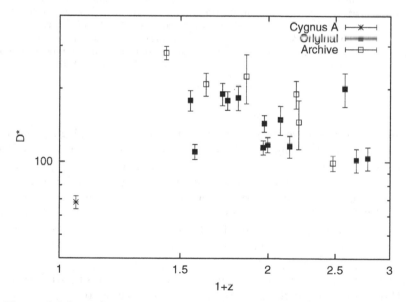

FIGURE 7. The model dependent, average size D_* the source would have if it could be observed over its entire lifetime computed assuming $\beta = 1.75$, $b = 0.25$, $\Omega_m = 0.1$, and $\Omega_\Lambda = 0$; b is the offset of the bridge magnetic field strength from the minimum energy value. Note that $D_* \propto (a_o r)^{-0.6}$.

that the shapes of the best-fitting curves to $D_*(z)$ and $<D>(z)$ match. That is why the method is independent of the local distance scale and of local sources. Differences between cosmological models begin to be obvious at redshifts greater than about 0.5; As shown in Figure 6, the slope of the line that describes $<D>(z)$ steepens as the

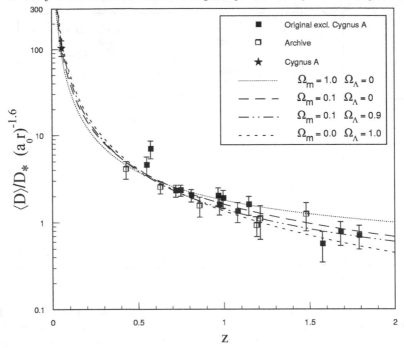

FIGURE 8. The quantity $(<D>/D_*)(a_o r)^{-1.6}$, computed assuming $\beta = 1.75$ and $b = 0.25$, where b parameterizes the offset from minimum energy conditions. Best fit models for different choices of Ω_m and Ω_Λ are shown, where sources with redshifts less than 0.3 have been **excluded** from the fits. Note that the fits are not normalized using local or low-redshift sources, yet all of the fits go right through the low-redshift point at a redshift of about 0.06 (Cygnus A) and $<D>/D_*(a_o r)^{-1.6} \sim 100$. The normalization is allowed to float for each fit.

cosmological parameters move from being dominated by a cosmological constant to an open universe to a matter-dominated $\Omega_m = 1$ universe; this arises because $<D> \propto (a_o r)$. However, the slope of the line that describes $D_*(z)$ begins at its steepest and becomes less steep as the cosmological parameters move from being dominated by a cosmological constant to an open universe to a matter-dominated $\Omega_m = 1$ universe; this arises because $<D> \propto (a_o r)^{-0.6}$. Thus, the power of the method lies in the fact that only for the correct choice of cosmological parameters will the shapes of $D_*(z)$ and $<D>(z)$ be the same. The normalization of the fits is allowed to float; that is $<D>/D_* = $ constant, and the constant is an output of the fit.

The ratio of $<D>/D_*$ is shown in Figure 8, where the ratio is multiplied by $(a_o r)^{-1.6}$ to remove the dependence of the ratio on cosmological parameters; a value of $\beta = 1.75$ was adopted for illustrative purposes. The fits shown are obtained using only radio data with redshifts greater than 0.3, thus, this is independent of the local distance scale and the properties of local sources. The best fitting lines pass right through the data point at a redshift of about 0.06 (Cygnus A) and $<D>/D_*(a_o r)^{-1.6} = 100$, which shows that the method is working quite well.

Evolution of source properties with redshift are much less of a concern for the radio galaxy method than for methods that rely upon the properties of local sources because the method relies on a comparison of individual source properties with the properties of the parent population.

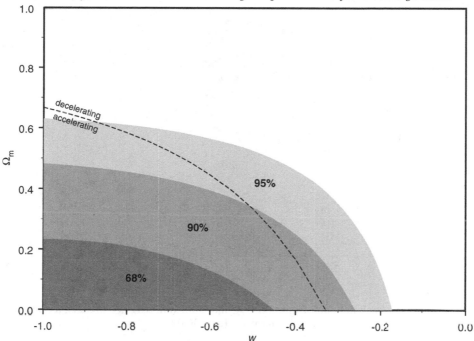

FIGURE 9. One-dimensional constraints on Ω_m and w obtained using D_* for 20 FRIIb radio galaxies, and assuming zero space curvature (see DG02).

4. Quintessence in a spatially flat universe

The application of FRIIb radio galaxies to constrain the current value of the normalized mean mass-energy density Ω_Q and (time-independent) equation of state $w - \Gamma/\rho$ of quintessence in a spatially flat universe, $\Omega_Q = 1 - \Omega_m$, is presented by DG02. The results are summarized here.

A universe with two primary components at the present epoch (non-relativistic matter and quintessence) will be accelerating in its expansion when, as seen in (DG02)

$$1 + 3w(1 - \Omega_m) < 0 \ . \tag{4.1}$$

Figure 9 shows the one-dimensional constraints obtained using a sample of 20 FRIIb radio galaxies, which are a subset of the 70 FRIIb radio galaxies in the parent population. Most of the sources in the sample have redshifts between 0.5 and 2, and identical results obtain with and without the lowest redshift bin; again, the method is independent of the local distance scale and of the properties of local sources. Clearly, FRIIb radio galaxies indicate that the universe is likely to be accelerating in its expansion at the present time; this is about a 2 σ result.

In any method, it is important to consider any possible covariance between constraints placed on different sets of parameters. For the FRIIb radio galaxy method, it is important to determine if there is any covariance between the global cosmological parameter Ω_m and the model parameter β. Figure 10 shows that there is no covariance between these parameters.

It is also important to insure that there is no covariance between the equation of state of quintessence w and the model parameter β. Figure 11 shows that there is no covariance between these parameters.

Clearly, a cosmological constant, which has $w = -1$, is consistent with the data.

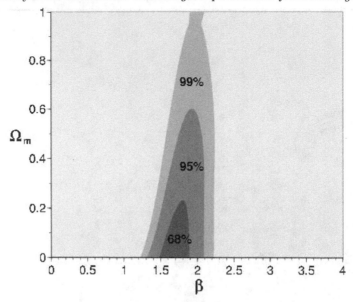

FIGURE 10. One-dimensional constraints on Ω_m and β for quintessence models, assuming zero space curvature (see DG02).

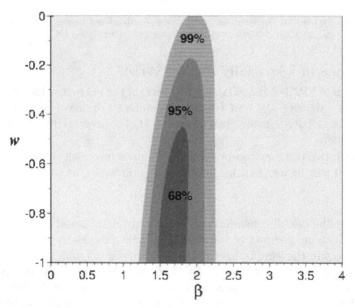

FIGURE 11. One-dimensional constraints on w and β for quintessence models, assuming zero space curvature (see DG02).

FRIIb radio galaxies can also be used to constrain global cosmological parameters allowing for space curvature, non-relativistic matter, and a cosmological constant (GDW00). The results indicate that Ω_m must be less than one, and $\Omega_m = 1$ is ruled out at about 99% confidence. Figures 12 and 13 show that there is no covariance between the model parameter β and the cosmological parameter Ω_m allowing for a cosmological constant, or space curvature.

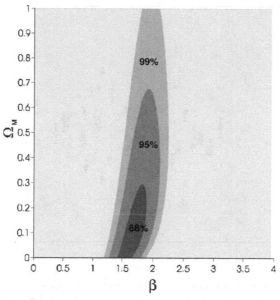

FIGURE 12. One-dimensional constraints on Ω_m and β assuming zero space curvature and allowing for a cosmological constant (see GDW00).

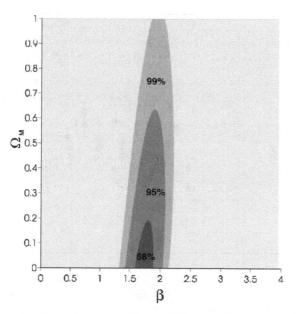

FIGURE 13. One-dimensional constraints on Ω_m and β assuming no cosmological constant and allowing for space curvature (see GDW00).

5. Implications for the beam power–total energy connection

Studies of FRIIb radio galaxies indicate that the total time t_* for which the central black hole produces collimated jets with beam power L_j that ultimately power the strong shocks near the hotspots and lobes of a given source are described by

$$t_* \propto L_j^{-\beta/3} \quad . \tag{5.1}$$

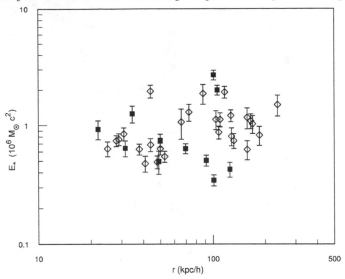

FIGURE 14. The total energy processed through each beam of each source as a function of core-hotspot separation; open symbols represent the 14 radio galaxies in the original sample (see GD98), and the filled symbols represent the 6 radio galaxies obtained from the VLA archive (see GDW00). Most sources have 2 data points, one from each side of the source.

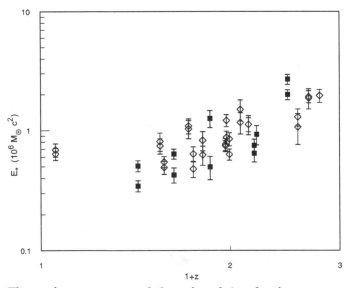

FIGURE 15. The total energy processed through each jet of each source as a function of redshift. The symbols are the same as in Figure 14.

The total energy $E_* = L_j t_* \propto L_j^{1-\beta/3}$, and the beam power is related to the total energy available to power the outflow

$$L_j \propto E_*^{3/(3-\beta)} \quad . \tag{5.2}$$

Thus, $L_j \propto E_*^{2 \ to \ 3}$ for $\beta = 1.5$ to 2; recall that $\beta = 1.75 \pm 0.25$.

An analysis of the implications for models of jet production is given by DG02, and is briefly summarized here. If the jets are powered by the electromagnetic extraction of the rotational energy of a spinning black hole (Blandford 1990), then equations 5.1 and 5.2 imply that the magnetic field B must be related to the black hole mass M, the spin

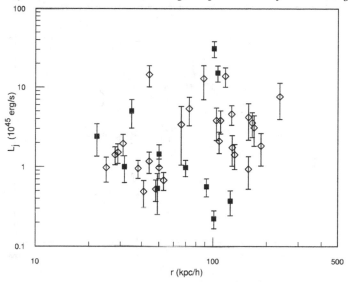

FIGURE 16. The beam power processed through each side of each source as a function of core-hotspot separation. The symbols are the same as in Figure 14.

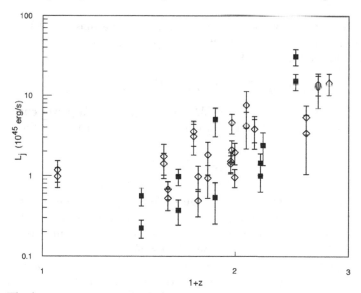

FIGURE 17. The beam power processed through each side of each source as a function of redshift. The symbols are the same as in Figure 14.

angular momentum S per unit mass $a = S/(Mc)$, and the effective size of the black hole $m = GM/c^2$, via the relation

$$B \propto M^{(2\beta-3)/2(3-\beta)}(a/m)^{\beta/(3-\beta)} \; , \qquad (5.3)$$

as detailed by DG02. Equation 5.3 takes a particularly simple form for $\beta = 1.5$; in this case, $B \propto (a/m)$. For a value of β of 1.75, equation 5.3 implies that $B \propto M^{0.2}(a/m)^{1.4}$, and for $\beta = 2$, $B \propto M^{1/2}(a/m)^2$. Specific values for B, L_j, and E_* are described below, where this model is discussed further.

If the jets are powered by radiation that is Eddington limited so $L_j = \eta_L L_E \propto \eta_L M$, where L_E is the Eddington luminosity, η_L is the efficiency with which the radiant lumi-

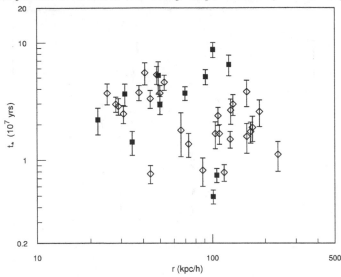

FIGURE 18. The total jet lifetime for each side of each source as a function of core-hotspot separation. The symbols are the same as in Figure 14.

nosity is converted into beam power, M is the black hole mass, and the accreted mass M_{acc} is converted to energy with an efficiency η_E so that $E_* = \eta_E M_{\mathrm{acc}} \, c^2$ (e.g. Krolik 1999), then equation 5.2 implies that

$$\eta_L \propto \eta_E^{3/(3-\beta)} (M/M_{\mathrm{acc}})^{-3/(3-\beta)} M^{\beta/(3-\beta)} \quad . \tag{5.4}$$

For $\beta = 1.5$, this reduces to $\eta_L \propto \eta_E^2 (M/M_{\mathrm{acc}})^{-2} M$. The beam power could be as high as the Eddington luminosity, or could be a constant fraction of the Eddington luminosity, in which case

$$\eta_E \propto (M/M_{\mathrm{acc}}) M^{-\beta/3} \quad , \tag{5.5}$$

or $\eta_E \propto M^{-1/2}$ for $\beta = 1.5$ and $M_{\mathrm{acc}} \sim M$. The total energy that will be processed by each source through each of its two large-scale jets over its lifetime E_* is shown as a function of core-hotspot separation in Figure 14 and as a function of redshift in Figure 15.

The beam powers of the sources are shown in Figure 16 as a function of core-hotspot separation, and in Figure 17 as a function of redshift.

Clearly, the beam power increases with redshift. However, if the beam power is related to the Eddington luminosity, then it is proportional to the black hole mass, which is expected to increase with time, or decrease with redshift. In addition, the beam powers are typically about 10^{45} erg s^{-1}, implying a black hole mass of only about 10^7 M$_\odot$, which is smaller than the black hole masses associated with galaxies in the cores of clusters of galaxies. The scaling of the beam power, total energy, and total source lifetime as given above, and can be normalized using the fact that when the beam power is $L_j \sim 10^{45}$ erg s^{-1}, the total active lifetime is $t_* \sim 10^7$ yr, and the total energy processed through the jet is $\sim 5 \times 10^5$ c^2 M$_\odot$ (see DG02, or Figures 14, 16, and 18).

Since the sources seem to be preferentially located near the cores of clusters or protoclusters of galaxies (see Wellman, Daly, & Wan 1997a [WDW97a] and WDW97b), a model in which a massive black hole acquires substantial rotational energy after coalescing with another massive black hole, such as that of Wilson & Colbert (1995) is quite attractive. A timescale can be constructed by dividing E_* by L_{EM}; using equations (3.38) and (3.39) from Blandford (1990), the timescale of the outflow is of order $10^9/(M_8 B_4^2)$ yr,

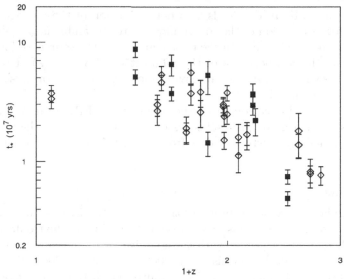

FIGURE 19. The total jet lifetime for each side of each source as a function of redshift. The symbols are the same as in Figure 14.

where M_8 is the mass of the black hole in units of 10^8 $M_\odot$, and B_4 is the magnetic field strength in units of 10^4 G. A total lifetime of about 10^7 yr results for a black hole mass of about 10^9 $M_\odot$ and a field strength of about 3×10^4 G. These values would lead to beam powers of about 10^{45} erg s^{-1} and total energies of about 10^{60} erg, equivalent to a rest mass of about 5×10^5 $M_\odot$, for $(a/m) \sim 1/30$, which are reasonable numbers, and which agree with the results obtained empirically.

For comparison, the total lifetimes of the 20 sources studied here are shown in Figure 18 as a function of core-hotspot separation, and in Figure 19 as a function of redshift. Clearly, high-redshift sources have shorter total lifetimes.

6. FRIIb radio sources provide probes of evolution of structure

The structure of the radio bridges of FRIIb sources can be used to determine the ambient gas pressure, density, temperature, and the overall Mach number of the forward region of the source, as detailed in a series of papers (Wan & Daly 1998; WDW97a,b; WDG00). The use of the properties of the radio bridge to determine the ambient gas pressure does not rely upon the assumption of minimum energy conditions, and it is independent of the rate of growth, or lobe propagation velocity, of the source, and hence is independent of any aging analysis applied to the radio bridge. The sources are in gaseous environments with pressures and composite pressure profiles like those found in clusters of galaxies, though some redshift evolution of the ambient gas pressure is suggested by the results obtained.

The ambient gas density may be obtained by studying the ram pressure confinement of the forward region of the radio source. This is done by combining the rate of growth of the source, determined using a spectral aging analysis across the radio bridge of the source, with the pressure near the end of the source, as done by Perley & Taylor (1990) for 3C295, Carilli et al. (1990) for Cygnus A, WDW97a for a sample of 14 radio galaxies and 8 radio loud quasars, and GDW00 for a sample of 6 radio galaxies. These studies indicate that the sources are in cluster-like gaseous environments, though some redshift evolution of the density and density profile is indicated by the study of WDW97b.

The shape of the radio bridge leads to a determination of the average Mach number with which the forward region of the source moves into the ambient gas (WDW97a). The Mach number can be combined with the rate of growth of the source, or lobe propagation velocity, to obtain the temperature of the ambient gas (WDW97a). This analysis also indicates that the sources are in gaseous environments with temperatures similar to those in present day cluster of galaxies.

Thus, there are some indications from the properties of FRIIb radio sources that they are in the cores of clusters or proto-clusters of galaxies, and that the properties of the gas in the cores of galaxy clusters evolves from a redshift of zero to a redshift of two.

7. Conclusions

FRIIb radio galaxies provide a probe of the coordinate distance to very high redshift sources, including sources with redshifts between one and two. The results obtained using FRIIb radio galaxies are consistent with those indicated by measurements of the cosmic microwave background (de Bernardis et al. 2000, Balbi et al. 2000, Bond et al. 2000), and type Ia supernovae (Riess et al. 1998, Perlmutter et al. 1999). The modified standard yardstick method, which uses FRIIb radio galaxies, is complementary to the modified standard candle method, which uses type Ia supernovae. Possible sources of error in each method are likely to be very different. Thus, the fact that they yield consistent results suggests that any errors not yet accounted for in either method must be small compared with errors that are currently known and accounted for.

FRIIb radio galaxies indicate that Ω_m must be low; $\Omega_m = 1$ is ruled out at about 99% confidence. Measurements of the cosmic microwave background radiation indicate that the universe has zero space curvature (e.g. Bond et al. 2000). In a universe with zero space curvature and quintessence with equation of state w, the modified standard yardstick method using FRIIb radio galaxies indicates that it is likely that the expansion of the universe is accelerating at present. The expansion rate of the universe will be accelerating when

$$1 + 3w(1 - \Omega_m) < 0 \ , \tag{7.1}$$

DG02. This line is drawn on Figure 9; the region below the line indicates that the universe is accelerating at present; those above the line indicate a decelerating universe. Radio galaxies alone indicate that the universe is accelerating in its expansion at the current epoch; this is an 84% confidence result.

The application of the modified standard yardstick method not only allows a determination of global cosmological parameters, it also allows a determination of the model parameter β; current results indicate that β is about 1.75 ± 0.25. The parameter β can be used to constrain models of energy extraction from the central massive object, presumed to be a massive black hole; these constraints are described by DG02 and are summarized in Section 5. The value of β determined empirically is expected/predicted in models where jet formation, power, and energy are related to the electromagnetic extraction of the rotational energy of a spinning black hole, and is consistent with models in which jet production is related to the Eddington luminosity of the black hole region.

Independent of the application of FRIIb radio galaxies to constrain cosmological parameters and models of energy extraction from massive black holes, FRIIb radio galaxies can also be used to study the gaseous environments of this type of AGN. The sources may be used to study the pressure, density, and temperature of the gas around them, and appear to be located in the cores of clusters or protoclusters of galaxies, as summarized in Section 6.

FRIIb radio sources appear to be governed by strong shock physics, and hence are in a regime that is easy to quantify. This fact, coupled with the interesting relation between the beam power and total energy that all of the sources seem to follow, makes them ideally suited to cosmological studies. In addition to their use as a modified standard yardstick, they also provide a probe of models of evolution of structure through their use to study the properties of the cores of clusters or proto-clusters of galaxies. The fact that they are located near the centers of clusters or proto-clusters of galaxies is probably related to the similar physical mechanism of jet formation and energy extraction.

It is a pleasure to thank Megan Donahue, Paddy Leahy, Chris O'Dea, Adam Reiss, and Max Tegmark for helpful comments and discussions. We are grateful to Lin Wan and Greg Wellman for their contributions to the study of FRIIb radio sources. This research was supported in part by National Young Investigator Award AST-0096077 from the US National Science Foundation, and by the Berks-Lehigh Valley College of Penn State University. Research at Rowan University was supported in part by the College of Liberal Arts and Sciences and National Science Foundation grant AST-9905652.

REFERENCES

BALBI, A. ET AL. 2000 *ApJ* **545**, L1.

BLANDFORD, R. D. 1990. In *Active Galactic Nuclei* (eds. T. J. L. Courvoisier & M. Mayor), p. 161. Springer-Verlag.

BOND ET AL. 2000 preprint (astro-ph/0011378).

CARILLI, C. L., PERLEY, R. A., DREHER, J. W., & LEAHY, J. P. 1991 *ApJ* **383**, 554.

DALY, R. A. 1994 *ApJ* **426**, 38.

DALY, R. A. 2002 *New Astr. Rev.* **46**, 47.

DALY, R. A. & GUERRA, E. J. 2002 *AJ*, **124**, 1831 [DG02].

DE BERNARDIS, P. ET AL. 2000 *Nature* **404**, 995.

GUERRA, E. J. 1997 *PhD Thesis*, Princeton University.

GUERRA, E. J. & DALY, R. A. 1998 *ApJ* **493**, 536 [GD98].

GUERRA, E. J., DALY, R. A., & WAN, L. 2000 *ApJ*, **544**, 659 [GDW00].

KROLIK, J. H. 1999 *Active Galactic Nuclei*, Princeton University Press.

PERLEY, R. A. & TAYLOR, G. B. 1991 *AJ* **101**, 1623.

PERLMUTTER ET AL. 1999 *ApJ* **517**, 565.

PERLMUTTER, W., TURNER, M. S., & WHITE, M. 1999 *Phys. Rev. Lett* 83, **No. 4**, 670.

RIESS ET AL. 1998 *AJ* **116**, 1009.

RIESS, A. G. 2000 *PASP* **112**, 1284.

TURNER, M. S. & WHITE, M. 1997 *Phys. Rev. D* 56, **No. 8**, 4439.

WANG, L., CALDWELL, R. R., OSTRIKER, J. P., & STEINHARDT, P. J. 2000 *ApJ* **530**, 17.

WAN, L. & DALY, R. A. 1998 *ApJ* **499**, 614.

WAN, L., DALY, R. A., & GUERRA, E. J. 2000 *ApJ* **544**, 671 [WDG00].

WELLMAN, G. F., DALY, R. A., & WAN, L. 1997a *ApJ* **480**, 79 [WDW97a].

WELLMAN, G. F., DALY, R. A., & WAN, L. 1997b *ApJ* **480**, 96 [WDW97b].

WILSON, A. S. & COLBERT, E. J. M. 1995 *ApJ* **438**, 62.

The mass density of the Universe

By NETA A. BAHCALL

Princeton University, Astrophysical Sciences Dept., Peyton Hall, Princeton, NJ 08544, USA

One of the most fundamental questions in cosmology is: How much matter is there in the Universe and how is it distributed? Here I review several independent measures—including those utilizing clusters of galaxies—that show that the mass-density of the Universe is only $\sim 20\%$ of the critical density. Recent measurements of the mass-to-light function—from galaxies, to groups, clusters, and superclusters—provide a powerful new measure of the universal density. The results reveal a low density of 0.16 ± 0.05 the critical density. The observations suggest that, on average, the mass distribution follows the light distribution on large scales. The results, combined with recent observations of high redshift supernovae and the spectrum of the CMB anisotropy, suggest a Universe that has low density ($\Omega_m \simeq 0.2$), is flat, and is dominated by dark energy.

1. Introduction

Theoretical arguments based on standard models of inflation, as well as on the demand of no "fine tuning" of cosmological parameters, predict a flat universe with the critical density needed to just halt its expansion (1.9×10^{-29} h^2 g cm^{-3}). Observations, however, reveal only a small fraction of the critical density, even when all the unseen dark matter in galaxy halos and clusters of galaxies is included. There is no reliable indication that the matter needed to close the universe does in fact exist. Here I review several independent observations of clusters of galaxies which indicate, independently, that the mass density of the universe is sub-critical. These observations include the mass and mass-to-light ratio of galaxies, clusters, and superclusters of galaxies, the high baryon fraction observed in clusters, and the evolution of the number density of massive clusters with time; the latter method provides a powerful measure not only of the mass-density of the universe but also the amplitude of the mass fluctuations. The three independent methods—all simple and robust—yield consistent results of a low-density universe with mass approximately tracing light on large scales. The results are consistent with those derived from the high redshift supernovae observations, the CMB anisotropy spectrum, and recent weak lensing observations on large scales.

2. Cluster dynamics and the Mass-to-Light Function

Rich clusters of galaxies are the most massive virialized objects known. Cluster masses can be directly and reliably determined using three independent methods: the motion of galaxies within clusters (Zwicky 1957, Bahcall 1977, Carlberg et al. 1996); the temperature of the hot intracluster gas (Jones et al. 1984, Sarazin 1986, Evrard 1996); and gravitational lensing distortions of background galaxies (Tyson et al. 1990, Kaiser et al. 1993, Smail et al. 1995, Colley et al. 1996). All three independent methods yield consistent cluster masses (typically within radii of ~ 1 Mpc), indicating that we can reliably determine cluster masses within the observed scatter ($\sim \pm 30\%$).

The simplest argument for a low density universe is based on summing up all the observed mass(associated with light to the largest possible scales) by utilizing the well-determined masses of clusters. The masses of rich clusters of galaxies range from $\sim 10^{14}$ to 10^{15} h^{-1} M$_\odot$ within 1.5 h^{-1} Mpc radius of the cluster center (where h = $H_0/100$ km s^{-1} Mpc^{-1} denotes Hubble's constant). When normalized by the cluster luminosity, a median mass-to-light ratio of M/L$_B$ $\simeq 300 \pm 100$h in solar units (M$_\odot$/L$_\odot$)

is observed for rich clusters (L_B is the total luminosity of the cluster in the blue band, corrected for internal and Galactic absorption; Bahcall et al. 1995, Carlberg et al. 1996). When integrated over the entire observed luminosity density of the universe, this mass-to-light ratio yields a mass density of $\Omega_m = \rho_m/\rho_{crit} \simeq 0.2 \pm 0.07$ (where ρ_{crit} is the critical density needed to close the universe). The inferred density assumes that all galaxies exhibit the same high M/L_B ratio as clusters, and that mass follows light on large scales. Thus, even if all galaxies have as much mass per unit luminosity as do massive clusters, the total mass of the universe is only $\sim 20\%$ of the critical density. If one insists on esthetic grounds that the universe has a critical density ($\Omega_m = 1$), then most of the mass of the universe has to be unassociated with galaxies (i.e. with light). On large scales ($\gtrsim 1.5$ h^{-1} Mpc) the mass has to reside in "voids" where there is no light. This would imply, for $\Omega_m = 1$, a large bias in the distribution of mass versus light, with mass distributed considerably more diffusely than light.

Is there a strong bias in the universe, with most of the dark matter residing on large scales, well beyond galaxies and clusters? Analysis of the mass-to-light ratio of galaxies, groups, and clusters by Bahcall et al. (1995) suggests that there is not a large bias. The study shows that the M/L_B ratio of galaxies increases with scale up to radii of $R \sim 0.2$ h^{-1} Mpc, due to very large dark halos around galaxies (Ostriker et al. 1974, Rubin 1993). The M/L ratio, however, appears to flatten and remain approximately constant for groups and rich clusters from scales of ~ 0.2 to at least 1.5 h^{-1} Mpc and even beyond (Figure 1). The flattening occurs at $M/L_B \simeq 200$–300 h, corresponding to $\Omega_m \simeq 0.2$. (An $M/L_B \sim 1400$ h is needed for a critical density universe, $\Omega_m = 1$.) This observation contradicts the classical belief that the relative amount of dark matter increases continuously with scale, possibly reaching $\Omega_m = 1$ on large scales. The available data suggest that most of the dark matter may be associated with very large dark halos of galaxies and that clusters do not contain a substantial amount of additional dark matter, other than that associated with (or torn-off from) the galaxy halos, plus the hot intracluster gas. This flattening of M/L with scale suggests that the relative amount of dark matter does not increase significantly with scale above ~ 0.2 h^{-1} Mpc. In that case, the mass density of the universe is low, $\Omega_m \sim 0.2$–0.3, with no significant bias (i.e. mass approximately following light on large scales).

The mass and mass-to-light ratio of a supercluster of galaxies, on a scale of ~ 6 h^{-1} Mpc, was recently measured using observations of weak gravitational lensing distortion of background galaxies (Kaiser et al. 2001). The results yield a supercluster mass-to-light ratio (on 6 h^{-1} Mpc scale) of $M/L_B = 280 \pm 40$ h, comparable to the mean value obtained for the three individual clusters that are members of this supercluster. These results provide a powerful confirmation of the suggested flattening of M/L_B (R) seen in Figure 1 (Bahcall et al. 1995, 1998).

Recently, Bahcall et al. (2000) used large-scale cosmological simulations to estimate the mass-to-light ratio of galaxy systems as a function of scale, and compare the results with observations of galaxies, groups, clusters, and superclusters of galaxies. They find remarkably good agreement between observations and simulations (Figure 1). Specifically, they find that the simulated mass-to-light ratio increases with scale on small scales and flattens to a constant value on large scales, as suggested by observations. The results show that while mass typically follows light on large scales, high overdensity regions—such as rich clusters and superclusters of galaxies—exhibit higher M/L_B values than average, while low density regions exhibit lower M/L_B values; high density regions are thus antibiased in M/L_B, with mass more strongly concentrated than blue light. The M/L_B antibias is mainly due to the relatively old age of the high density regions, where light has declined significantly since their early formation time, especially in the blue band

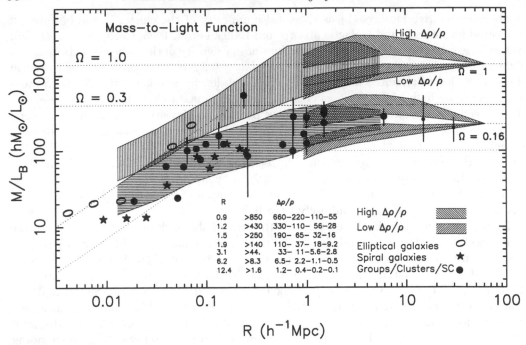

FIGURE 1. The mass-to-light function of galaxy systems from observations (Bahcall et al. 1995) and simulations (Bahcall et al. 2000). The observations are presented by the data points for medians of galaxies, groups, clusters, and a supercluster. The simulation results (for cold-dark-matter models) are presented by the shaded bands for $\Omega_m = 1$ and 0.16 (our best fit value). On scales > 1 Mpc, the simulation results for both high- and low-density regions are presented (where these correspond roughly to the overdensities of rich-clusters and groups, respectively). See Bahcall et al. (2000) for more details.

which traces recent star formation. Comparing the simulated results with observations, Bahcall et al. (2000) place a powerful constraint on the mass density of the universe; using, for the first time, the entire observed mass-to-light function, from galaxies to superclusters, they find

$$\Omega_m = 0.16 \pm 0.05 \quad . \tag{2.1}$$

3. Baryons in clusters

Clusters contain many baryons, observed as gas and stars. Within 1.5 h^{-1} Mpc of a rich cluster, the X-ray emitting gas contributes $\sim 7\ h^{-1.5}\%$ of the cluster virial mass (White et al. 1993, 1995, Lubin et al. 1996). Stars in the predominately early type cluster galaxies contribute another $\sim 3\%$. The baryon fraction observed in clusters is thus:

$$\Omega_b/\Omega_m \gtrsim 0.07\ h^{-1.5} + 0.03 \quad . \tag{3.1}$$

Standard Big Bang nucleosynthesis limits the baryon density of the universe to (Walker et al. 1991, Tytler et al. 1996):

$$\Omega_b \simeq 0.02\ h^{-2} \quad . \tag{3.2}$$

These facts suggest that the baryon fraction observed in rich clusters (Eq. 2) exceeds that of an $\Omega_m = 1$ universe ($\Omega_b/(\Omega_m = 1) \simeq 0.02\ h^{-2}$; Eq. 3) by a factor of $\gtrsim 3$ (for $h \simeq 0.5$). Since detailed hydrodynamic simulations (White et al. 1993, Lubin et al. 1996) show that baryons do not segregate into rich clusters, the above results imply that either the mean density of the universe is lower than the critical density by a factor of $\gtrsim 3$, or that the baryon density is much larger than predicted by nucleosynthesis. The observed high baryonic mass fraction in clusters, combined with the nucleosynthesis limit, indicate (for $h \simeq 0.65 \pm 0.1$):

$$\Omega_m \lesssim 0.3 \pm 0.05 \ . \tag{3.3}$$

This upper limit on Ω_m is a simple model-independent and thus powerful constraint: a critical density universe is inconsistent with the high baryon fraction observed in clusters. Observations of the Sunyaev-Zeldovich effect in clusters yield the same result (Carlstrom et al. 2001).

4. Evolution of cluster abundance

The observed present-day abundance of rich clusters of galaxies places a strong constraint on cosmology: $\sigma_8 \Omega_m^{0.5} \simeq 0.5$, where σ_8 is the *rms* mass fluctuations on 8 h^{-1} Mpc scale, and Ω_m is the present cosmological density parameter (Bahcall et al. 1992, White et al. 1993, Eke et al. 1996, Viana et al. 1996, Kitayama et al. 1996, Pen 1998). This constraint is degenerate in Ω_m and σ_8; models with $\Omega_m = 1$, $\sigma_8 \sim 0.5$ are indistinguishable from models with $\Omega_m \sim 0.25$, $\sigma_8 \sim 1$. (A $\sigma_8 \simeq 1$ universe is unbiased, with mass following light on large scales since galaxies (light) exhibits σ_8 (galaxies) $\simeq 1$; $\sigma_8 \simeq 0.5$ implies a mass distribution wider than light).

The *evolution* of cluster abundance with redshift, especially for massive clusters, breaks the degeneracy between Ω_m and σ_8 (Peebles et al. 1989, Eke et al. 1996, Viana et al. 1996, Oukbir et al. 1992, 1997, Carlberg et al. 1997, Bahcall et al. 1997, Fan et al. 1997, Henry 1997, Bahcall et al. 1998). The evolution of high mass clusters is strong in $\Omega_m = 1$, low-σ_8 (biased) Gaussian models, where only a very low cluster abundance is expected at $z > 0.5$. Conversely, the evolution rate in low-Ω_m, high-σ_8 models is mild and the cluster abundance at $z > 0.5$ is much higher than in $\Omega_m = 1$ models.

In low-density models, density fluctuations evolve and freeze out at early times, thus producing only relatively little evolution at recent times ($z \lesssim 1$). In an $\Omega_m = 1$ universe, the fluctuations start growing more recently thereby producing strong evolution in recent times; a large increase in the abundance of massive clusters is expected from $z \sim 1$ to $z \sim 0$. Bahcall et al. (1997) show that the evolution is so strong in $\Omega_m = 1$ models that finding even a few Coma-like clusters at $z > 0.5$ over $\sim 10^3$ deg^2 of sky contradicts an $\Omega_m = 1$ model where only $\sim 10^{-2}$ such clusters would be expected (when normalized to the present-day cluster abundance).

The evolutionary effects increase with cluster mass and with redshift. The existence of the three most massive clusters observed so far at $z \sim 0.5$–0.9 places the strongest constraint yet on Ω_m and σ_8. These clusters (MS0016+16 at $z = 0.55$, MS0451$-$03 at $z = 0.54$, and MS1054$-$03 at $z = 0.83$, from the Extended Medium Sensitivity Survey, EMSS (Henry et al. 1992, Luppino et al. 1995), are nearly twice as massive as the Coma cluster, and have reliably measured masses (including gravitational lensing masses, temperatures, and velocity dispersions; (Bahcall et al. 1998, Smail et al. 1995, Luppino et al. 1997, Mushotsky et al. 1997, Donahue et al. 1999). These clusters posses the highest masses ($\gtrsim 8 \times 10^{14}$ h^{-1} M$_\odot$ within 1.5 h^{-1} commoving Mpc radius), the highest velocity dispersions ($\gtrsim 1200$ km s^{-1}), and the highest temperatures ($\gtrsim 8$ keV) in the $z > 0.5$ EMSS survey. The existence of these three massive distant clusters, even just the existence of the single observed cluster at $z = 0.83$, rules out Gaussian $\Omega_m = 1$ models for which only $\sim 10^{-5}$ $z \sim 0.8$ clusters are expected instead of the 1 cluster observed (or $\sim 10^{-3}$ $z > 0.5$ clusters expected instead of the 3 observed; Bahcall et al. 1998, Donahue

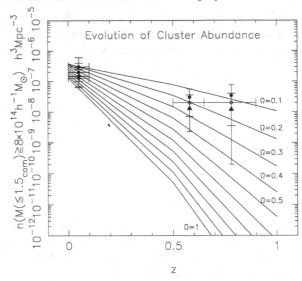

FIGURE 2. Evolution of the cluster abundance with redshift for massive clusters (with mass $> 8 \times 10^{14}$ h^{-1} $M_\odot$ within a comoving radius of 1.5 h^{-1} Mpc; from Bahcall & Fan 1998.) The data points represent the observational data (see text), and the curves represent the expected cluster abundance for different Ω_m values (based on the Press-Schechter method). Similar results are obtained from direct simulations (Bode et al. 2001).

& Voit 1999). Figure 2 compares the observed versus expected evolution of such massive clusters.

The data provide powerful constraints on Ω_m and σ_8: $\Omega_m = 0.2^{+0.15}_{-0.1}$ and $\sigma_8 = 1.2 \pm 0.3$ (68% confidence level; Bahcall et al. 1998). The high σ_8 value for the mean mass fluctuations indicates a nearly unbiased universe, with mass approximately tracing light on large scales. This conclusion is consistent with the suggested flattening of the observed M/L ratio on large scales (Figure 1).

5. Summary

We have shown that several independent observations of clusters of galaxies all indicate that the mass-density of the universe is sub-critical: $\Omega_m \simeq 0.2 \pm 0.1$. A summary of the results is highlighted below.

(*a*) The mass-to-light function of galaxies, groups, clusters and superclusters of galaxies yields a tight constraint; $\Omega_m = 0.16 \pm 0.05$.

(*b*) The high baryon fraction observed in clusters of galaxies suggests $\Omega_m \lesssim 0.3 \pm 0.05$.

(*c*) The weak evolution of the observed cluster abundance to $z \sim 1$ provides an independent estimate: $\Omega_m \simeq 0.2^{+0.15}_{-0.1}$, valid for any Gaussian models. An $\Omega_m = 1$ Gaussian universe is ruled out as a $\lesssim 10^{-6}$ probability by the cluster evolution results.

(*d*) All the above-described independent measures are consistent with each other and indicate a low-density universe with $\Omega_m \simeq 0.2 \pm 0.1$.

(*e*) The above results are consistent with those derived from weak lensing observations on large scales, from high redshift supernovae observations (assuming a flat universe) and from the recent CMB observations (indicating a flat universe; Lange et al. 2000, Netterfield et al. 2001, Halverson et al. 2001, Lee et al. 2001). Combining the above results of clusters, SNe, and CMB in the Cosmic Triangle (Bahcall et al. 1999), we find a universe that is lightweight ($\Omega_m = 0.2$) and flat.

REFERENCES

BAHCALL, N. A. 1977 *ARA&A* **15**, 505.

BAHCALL, N. A. & CEN, R. 1992 *ApJ* **398**, L81.

BAHCALL, N. A., CEN, R., DAVÉ, R., OSTRIKER, J. P., & YU, Q. 2000 *ApJ* **541**, 1.

BAHCALL, N. A. & FAN, X. 1998 *Proc. Nat. Acad. Sci. USA* **95**, 5956.

BAHCALL, N. A. & FAN, X. 1998 *ApJ* **504**, 1.

BAHCALL, N. A., FAN, X., & CEN, R. 1997 *ApJ* **485**, L53.

BAHCALL, N. A., LUBIN, L., & DORMAN, V. 1995 *ApJ* **447**, L81.

BAHCALL, N. A., OSTRIKER, J. P., PERLMUTTER, S., STEINHARDT, P. 1999 *Science* **241**, 1481.

BODE, P., BAHCALL, N. A., FORD, E. OSTRIKER, J. O. 2001 *ApJ* **551**, 15.

CARLBERG, R. G., ET AL. 1996 *ApJ* 462, 32.

CARLBERG, R. G., MORRIS, S. M., YEE, H. K. C., & ELLINGSON, E. 1997 *ApJ* **479**, L19.

CARLSTROM, J., ET AL. 2001. In *IAP Conference 2000, Constructing the Universe with Clusters of Galaxies*, (eds. F. Durrett & D. Gerbal).

COLLEY, W. N., TYSON, J. A., & TURNER, E. L. 1996 *ApJ* 461, L83.

DONAHUE, M. & VOIT, G. M. 1999 *ApJ* **523**, L137.

EKE, V. R., COLE, S., & FRENK, C. S. 1996 *MNRAS* **282**, 263.

EVRARD, A. E., METZLER, C. A., & NAVARRO, J. F. 1996 *ApJ* **469**, 494.

FAN, X., BAHCALL, N. A., & CEN, R. 1997 *ApJ* **490**, L123.

HALVERSON, N. W., ET AL. 2001 *ApJ* submitted, astro-ph/0104460.

HENRY, J. P., ET AL. 1992 *ApJ* **386**, 408.

HENRY, J. P. 1997 *ApJ* **489**, L1.

JONES, C. & FORMAN, W. 1984 *ApJ* **276**, 38.

KAISER, N. & SQUIRES, G. 1993 *ApJ* **404**, 441.

KAISER, N., ET AL. 2001 *ApJ* in press, astro-ph/9809268.

KITAYAMA, T. & SUTO, Y. 1996 *ApJ* **469**, 480.

LANGE, A., ET AL. 2000 astro-ph/0005004.

LEE, A. T., ET AL. 2001 astro-ph/0104459.

LUBIN, L., CEN, R., BAHCALL, N. A., & OSTRIKER, J. P. 1996 *ApJ* **460**, 10.

LUPPINO, G. A. & GIOIA, I. M. 1995 *ApJ* **445**, L77.

LUPPINO, G. A. & KAISER, N. 1997 *ApJ* **475**, 20.

MUSHOTSKY, R. & SCHARF, C. A. 1997 *ApJ* **482**, L13.

NETTERFIELD, C. B., ET AL. 2001 astro-ph/0104460.

OSTRIKER, J. P., PEEBLES, P. J. E., & YAHIL, A. 1974 *ApJ* **193**, L1.

OUKBIR, J. & BLANCHARD, A. 1992 *A&A* **262**, L21.

OUKBIR, J. & BLANCHARD, A. 1997 *A&A* **317**, 1.

PEEBLES, P. J. E., DALY, R. A., & JUSZKIEWICZ, R. 1989 *ApJ* **347**, 563.

PEN, U.-L. 1998, *ApJ* **498**, 60.

RUBIN, V. C. 1993, *Proc. Natl. Acad. Sci. USA* **90**, 4814.

SARAZIN, C. L. 1996, *Rev. Mod. Phys.* **58**, 1.

SMAIL, I., ELLIS, R. S., FITCHETT, M. J., & EDGE, A. C. 1995 *MNRAS* **273**, 277.

TYSON, J. A., WENK, R. A., & VALDES, F. 1990 *ApJ* **349**, L1.

TYTLER, D., FAN, X.-M., & BURLES, S. 1996 *Nature* **381**, 207.

VIANA, P. P. & LIDDLE, A. R. 1996 *MNRAS* **281**, 323.

WALKER, T. P., ET AL. 1991 *ApJ* **376**, 51.

WHITE, D. & FABIAN, A. 1995 *MNRAS* **272**, 72.

WHITE, S. D. M., EFSTATHIOU, G., & FRENK, C. S. 1993 *MNRAS* **262**, 1023.

WHITE, S. D. M., NAVARRO, J. F., EVRARD, A., & FRENK, C. S. 1993 *Nature* **366**, 429.

ZWICKY, F. 1957 *Morphological Astronomy* (Berlin: Springer-Verlag).

Growth of structure in the Universe

By J. A. PEACOCK

Institute for Astronomy, University of Edinburgh, Royal Observatory, Edinburgh EH9 3HJ, UK

The simplest models for the formation of large-scale structure are reviewed. On the assumption that the dark matter is cold and collisionless, LSS data are able to measure the total amount of matter, together with the baryon fraction and the spectral index of primordial fluctuations. There are degeneracies between these parameters, but these are broken by the addition of extra information such as CMB fluctuation data. The CDM models are confronted with recent data, especially the 2dF Galaxy Redshift Survey, which was the first to measure more than 100,000 redshifts. The 2dFGRS power spectrum is measured to $\lesssim 10\%$ accuracy for $k > 0.02\,h\,\mathrm{Mpc}^{-1}$, and is well fitted by a CDM model with $\Omega_m h = 0.20 \pm 0.03$ and a baryon fraction of 0.15 ± 0.07. In combination with CMB data, a flat universe with $\Omega_m \simeq 0.3$ is strongly favored. In order to use LSS data in this way, an understanding of galaxy bias is required. A recent approach to bias, known as the 'halo model' allows important insights into this phenomenon, and gives a calculation of the extent to which bias can depend on scale.

---

1. Structure formation in the CDM model

The origin and formation of large-scale structure in cosmology is a key problem that has generated much work over the years. Out of all the models that have been proposed, this talk concentrates on the simplest: gravitational instability of small initial density fluctuations. Furthermore, it is assumed that the mass density is dominated by a collisionless component, so that we are left with a theory very like the Cold Dark Matter model. This does not mean that CDM is an untestable religion, to which cosmologists cling in the face of all evidence. However, it is the simplest model for structure formation, and thus should be tested thoroughly before we move on to more complex alternatives.

1.1. The CDM power spectrum

Suppose there existed some primordial power-law spectrum, written dimensionlessly as the logarithmic contribution to the fractional density variance, σ^2:

$$\Delta^2(k) = \frac{d\sigma^2}{d\ln k} \propto k^{3+n} \quad . \tag{1}$$

This undergoes linear growth

$$\delta_k(a) = \delta_k(a_0) \left[\frac{D(a)}{D(a_0)} \right] T_k \quad , \tag{2}$$

where the linear growth law is $D(a) = a\, g[\Omega(a)]$ in the matter era, and the growth suppression for low Ω is

$$g(\Omega) \simeq \Omega^{0.65} \ \text{(open)}$$
$$\simeq \Omega^{0.23} \ \text{(flat)} \tag{3}$$

The transfer function T_k depends on the dark-matter content as shown in Figure 1.

The state of the linear-theory spectrum after these modifications is illustrated in Figure 2. The primordial power-law spectrum is reduced at large k, by an amount that depends on both the quantity of dark matter and its nature. Generally the bend in the spectrum occurs near $1/k$ of order the horizon size at matter-radiation equality, $\propto (\Omega h^2)^{-1}$. For a pure CDM universe, with scale-invariant initial fluctuations ($n = 1$),

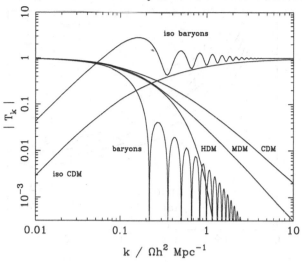

FIGURE 1. Transfer functions for various dark-matter models. The scaling with Ωh^2 is exact only for the zero-baryon models; the baryon results are scaled from the particular case $\Omega_B = 1$, $h = 1/2$.

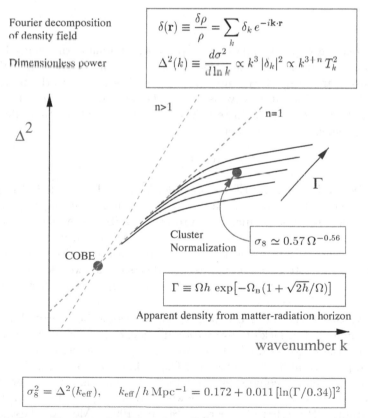

Fourier decomposition of density field

$$\delta(\mathbf{r}) \equiv \frac{\delta\rho}{\rho} = \sum_k \delta_k \, e^{-i\mathbf{k}\cdot\mathbf{r}}$$

Dimensionless power

$$\Delta^2(k) \equiv \frac{d\sigma^2}{d\ln k} \propto k^3 |\delta_k|^2 \propto k^{3+n} T_k^2$$

n>1

n=1

Δ^2

Γ

Cluster Normalization

$$\sigma_8 \simeq 0.57 \, \Omega^{-0.56}$$

COBE

$$\Gamma \equiv \Omega h \, \exp\left[-\Omega_B(1 + \sqrt{2h}/\Omega)\right]$$

Apparent density from matter-radiation horizon

wavenumber k

$$\sigma_8^2 = \Delta^2(k_{\text{eff}}), \qquad k_{\text{eff}}/h\,\text{Mpc}^{-1} = 0.172 + 0.011\,[\ln(\Gamma/0.34)]^2$$

FIGURE 2. This figure illustrates how the primordial power spectrum is modified as a function of density in a CDM model. For a given tilt, it is always possible to choose a density that satisfies both the COBE and cluster normalizations.

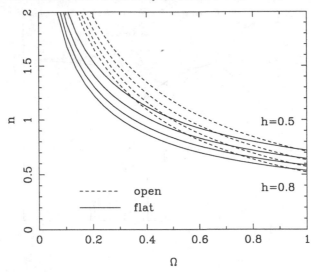

FIGURE 3. For 15% baryons, the value of n needed to reconcile COBE and the cluster normalization in CDM models. If $h = 0.7$, then $\Omega_m \simeq 0.35$ is required in a flat universe containing scale-invariant fluctuations.

the observed spectrum depends only on two parameters. One is the shape $\Gamma = \Omega h$, and the other is a normalization. This can be set at a number of points. The COBE normalization comes from large angle CMB anisotropics, and is sensitive to the power spectrum at $k \simeq 10^{-3}\,h\,\mathrm{Mpc}^{-1}$. The alternative is to set the normalization near the quasilinear scale, using the abundance of rich clusters. Many authors have tried this calculation, and there is good agreement on the answer:

$$\sigma_8 \simeq 0.5\,\Omega_m^{-0.6} \ . \tag{4}$$

(White, Efstathiou & Frenk 1993; Eke et al. 1996; Viana & Liddle 1996; Pierpaoli, Scott & White 2001; Viana et al. 2001; Seljak 2001; Reiprich & Böhringer 2001). In many ways, this is the most sensible normalization to use for LSS studies, since it does not rely on an extrapolation from larger scales. Within the CDM model, it is always possible to satisfy both these normalization constraints, by appropriate choice of Γ and n. This is illustrated in Figure 3. Note that vacuum energy affects the answer; for reasonable values of h and reasonable baryon content, flat models require $\Omega_m \simeq 0.3$, whereas open models require $\Omega_m \simeq 0.5$.

Figure 4 shows that rather large oscillatory features would be expected if the universe was baryon dominated. The lack of observational evidence for such features is one reason for believing that the universe might be dominated by collisionless nonbaryonic matter (consistent with primordial nucleosynthesis if $\Omega_m \gtrsim 0.1$). Nevertheless, baryonic fluctuations in the spectrum can become significant for high-precision measurements. Figure 4 shows that order 10% modulation of the power may be expected in realistic baryonic models (Eisenstein & Hu 1998; Goldberg & Strauss 1998). Most of these features are however removed by nonlinear evolution. The highest-k feature to survive is usually the second peak, which almost always lies near $k = 0.05\,\mathrm{Mpc}^{-1}$ (no h, for a change). This feature is relatively narrow, and can serve as a clear proof of the past existence of baryonic oscillations in forming the mass distribution (Meiksin, White & Peacock 1999).

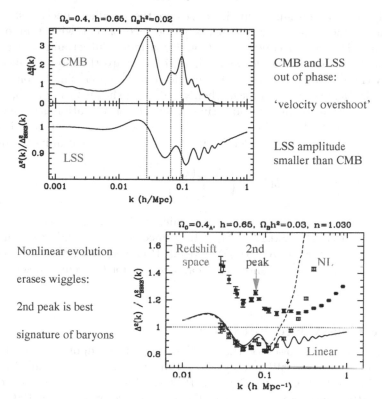

FIGURE 4. Baryonic fluctuations in the spectrum can become significant for high-precision measurements. Although such features are much less important in the density spectrum than in the CMB (first panel), the order 10% modulation of the power is potentially detectable. However, nonlinear evolution has the effect of damping all beyond the second peak. This second feature is relatively narrow, and can serve as a clear proof of the past existence of oscillations in the baryon-photon fluid (Meiksin, White & Peacock 1999).

1.2. *The need for scale-dependent bias*

An important cosmological goal is to compare the CDM predictions with what is measured in the galaxy distribution. To do this, it is necessary to allow for the effects of nonlinear evolution on the linear-theory CDM spectrum described above. This can be done analytically to high accuracy (e.g. Hamilton et al. 1991; Peacock & Dodds 1996).

Prior to about 2001, the best estimate of the galaxy power spectrum was that derived from angular clustering in the APM survey (Baugh & Efstathiou 1993, 1994; Maddox et al. 1996). The APM survey was generated from a catalogue of $\sim 10^6$ galaxies derived from UK Schmidt Telescope photographic plates scanned with the Cambridge Automatic Plate Measuring machine. The APM result has been investigated in detail by a number of authors (e.g. Gaztañaga & Baugh 1998; Eisenstein & Zaldarriaga 2001; Efstathiou & Moody 2001) and found to be robust, although there is some evidence that the true power uncertainty due cosmic variance may have been underestimated. One key advantage of the APM estimate of the power spectrum is that it is based on a deprojection of angular clustering, and is thus immune to the complicating effects of redshift-space distortions, discussed below.

On large scales, the general form of the CDM power spectrum seems to match the data from a number of galaxy tracers (Peacock & Dodds 1994). However, a number of authors have pointed out that, on small nonlinear scales, the detailed spectral shape inferred from APM and other galaxy data appears to be inconsistent with that of nonlinear evolution from CDM initial conditions. (e.g. Efstathiou, Sutherland & Maddox 1990; Klypin, Primack & Holtzman 1996; Peacock 1997). Perhaps the most detailed work was carried out by the VIRGO consortium, who carried out $N = 256^3$ simulations of a number of CDM models (Jenkins et al. 1998). Their results are shown in Figure 5, which gives the nonlinear power spectrum at various times (cluster normalization is chosen for $z = 0$) and contrasts this with the APM data. The lower small panels are the scale-dependent bias that would required if the model did in fact describe the real universe, defined as

$$b(k) \equiv \left(\frac{\Delta^2_{\text{gals}}(k)}{\Delta^2_{\text{mass}}} \right)^{1/2} . \tag{5}$$

In all cases, the required bias is non-monotonic; it rises at $k \gtrsim 5\,h^{-1}\,\text{Mpc}$, but also displays a bump around $k \simeq 0.1\,h^{-1}\,\text{Mpc}$. It seemed for some while that this behavior might constitute a fatal objection to CDM models. However, this is probably not so; at the end of this talk, a model is presented that can account for the required scale-dependent bias.

2. The 2dF Galaxy Redshift Survey

Despite the statistical power of the APM survey, it is important to study the 3D galaxy distribution via redshift surveys. These have progressed steadily from of order 10^3 redshifts in the CfA survey (Davis & Peebles 1983) to well over 10^4 redshifts in the 1990s (LCRS: Shectman et al. 1996; DUKST: Hoyle et al. 1999; PSCz: Saunders et al. 2000). The 2dF Galaxy Redshift Survey (2dFGRS) was designed to build on these previous studies, with the following main aims:

(*a*) To measure the galaxy power spectrum $P(k)$ on scales up to a few hundred Mpc, bridging the gap between the scales of nonlinear structure and measurements from the cosmic microwave background (CMB).

(*b*) To measure the redshift-space distortion of the large-scale clustering that results from the peculiar velocity field produced by the mass distribution.

(*c*) To measure higher-order clustering statistics in order to understand biased galaxy formation, and to test whether the galaxy distribution on large scales is a Gaussian random field.

2.1. *Survey design*

The survey is designed around the 2dF multi-fiber spectrograph on the Anglo-Australian Telescope, which is capable of observing up to 400 objects simultaneously over a 2 degree diameter field of view. Full details of the instrument and its performance are given in Lewis et al. (2001). See also http://www.aao.gov.au/2df/. For details of the current status of the 2dFGRS, see http://www.mso.anu.edu.au/2dFGRS. In particular, this site gives details of the 2dFGRS public release policy, in which approximately the first half of the survey data will be released by mid-2001, with the complete survey database to be made public by mid-2003.

The source catalogue for the survey is a revised and extended version of the APM galaxy catalogue (Maddox et al. 1990a,b,c). The extended version of the APM catalogue includes over 5 million galaxies down to $b_{\text{J}} = 20.5$ in both north and south Galactic hemispheres over a region of almost $10^4\ \text{deg}^2$ (bounded approximately by declination $\delta \leqslant +3$ and Galactic latitude $b \gtrsim 20$). This catalogue is based on Automated Plate Measuring

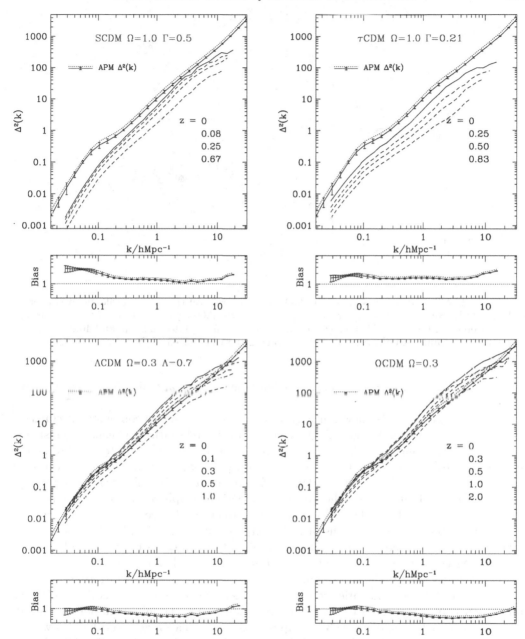

FIGURE 5. The nonlinear evolution of various CDM power spectra, as determined by the Virgo consortium (Jenkins et al. 1998). The dashed lines show the evolving spectra for the mass, which at no time match the shape of the APM data. This is expressed in the lower small panels as a scale-dependent bias at $z = 0$: $b^2(k) = P_{\mathrm{APM}}/P_{\mathrm{mass}}$.

machine (APM) scans of 390 plates from the UK Schmidt Telescope (UKST) Southern Sky Survey. The b_{J} magnitude system for the Southern Sky Survey is defined by the response of Kodak IIIaJ emulsion in combination with a GG395 filter. The photometry of the catalogue is calibrated with numerous CCD sequences and has a precision of approximately 0.2 mag for galaxies with $b_{\mathrm{J}} = 17$–19.5. The star-galaxy separation is as

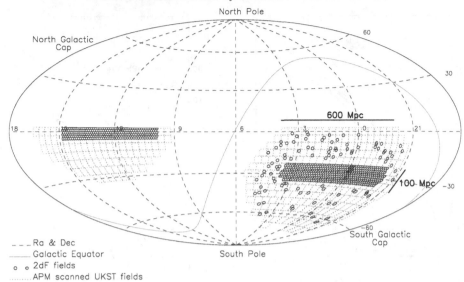

FIGURE 6. The 2dFGRS fields (small circles) superimposed on the APM catalogue area (dotted outlines of Sky Survey plates). There are approximately 140,000 galaxies in the $75° \times 15°$ southern strip centered on the SGP, 70,000 galaxies in the $75° \times 7.5°$ equatorial strip, and 40,000 galaxies in the 100 randomly-distributed 2dF fields covering the whole area of the APM catalogue in the south.

described in Maddox et al. (1990b), supplemented by visual validation of each galaxy image.

The survey geometry is shown in Figure 6, and consists of two contiguous declination strips, plus 100 random 2-degree fields. One strip is in the southern Galactic hemisphere and covers approximately $75° \times 15°$ centered close to the SGP at $(\alpha, \delta)=(01^h, -30)$; the other strip is in the northern Galactic hemisphere and covers $75° \times 7.5°$ centered at $(\alpha, \delta)=(12.5^h, +00)$. The 100 random fields are spread uniformly over the 7000 deg^2 region of the APM catalogue in the southern Galactic hemisphere. At the median redshift of the survey ($\bar{z} = 0.11$), $100 \, h^{-1}$ Mpc subtends about 20 degrees, so the two strips are $375 \, h^{-1}$ Mpc long and have widths of $75 \, h^{-1}$ Mpc (south) and $37.5 \, h^{-1}$ Mpc (north).

The sample is limited to be brighter than an extinction-corrected magnitude of $b_{\rm J} = 19.45$ (using the extinction maps of Schlegel et al. 1998). This limit gives a good match between the density on the sky of galaxies and 2dF fibers. Due to clustering, however, the number in a given field varies considerably. To make efficient use of 2dF, an adaptive tiling algorithm is employed to cover the survey area with the minimum number of 2dF fields. This algorithm achieves a 93% sampling rate with on average fewer than 5% wasted fibers per field. Over the whole area of the survey there are in excess of 250,000 galaxies.

2.2. Survey status

By the end of 2001, observations had been made of 866 fields, yielding redshifts and identifications for 224,851 galaxies, 13630 stars and 176 QSOs, at an overall completeness of 93%. The galaxy redshifts are assigned a quality flag from 1 to 5, where the probability of error is highest at low Q. Most analyses are restricted to $Q \geqslant 3$ galaxies, of which there are currently 213,703. Data-taking will continue in 2002, but concentrating on increasing the completeness of the existing survey zone; the total number of $Q \geqslant 3$ galaxies will probably asymptote to about 230,000. Figure 7 shows the projection of the galaxies in the northern and southern strips onto (α, z) slices. The main points to note are the level

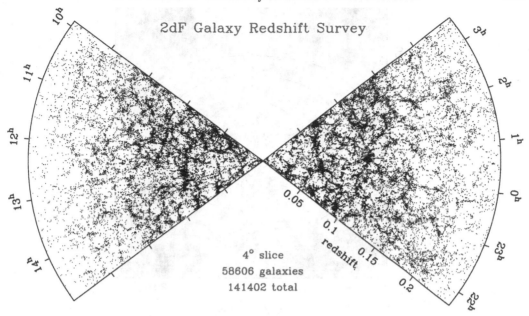

FIGURE 7. The distribution of galaxies in part of the 2dFGRS, drawn from a total of 141,402 galaxies: slices 4° thick, centered at declination −2.5° in the NGP and −27.5° in the SGP. Not all 2dF fields within the slice have been observed at this stage, hence there are weak variations of the density of sampling as a function of right ascension. To minimize such features, the slice thickness increases to 7.5° between right ascension 13.1^h and 13.4^h. This image reveals a wealth of detail, including linear supercluster features, often nearly perpendicular to the line of sight. The interesting question to settle statistically is whether such transverse features have been enhanced by infall velocities.

of detail apparent in the map and the slight variations in density with R.A. due to the varying field coverage along the strips.

The adaptive tiling algorithm is efficient, and yields uniform sampling in the final survey. However, at this intermediate stage, missing overlaps mean that the sampling fraction has large fluctuations. This variable sampling makes quantification of the large scale structure more difficult, and limits any analysis requiring relatively uniform contiguous areas. However, the effective survey 'mask' can be measured precisely enough that it can be allowed for in low-order analyses of the galaxy distribution.

3. Redshift-space correlations

The simplest statistic for studying clustering in the galaxy distribution is the two-point correlation function, $\xi(\sigma, \pi)$. This measures the excess probability over random of finding a pair of galaxies with a separation in the plane of the sky σ and a line-of-sight separation π. Because the radial separation in redshift space includes the peculiar velocity as well as the spatial separation, $\xi(\sigma, \pi)$ will be anisotropic. On small scales the correlation function is extended in the radial direction due to the large peculiar velocities in non-linear structures such as groups and clusters—this is the well-known 'Finger-of-God' effect. On large scales it is compressed in the radial direction due to the coherent infall of galaxies onto mass concentrations—the Kaiser effect (Kaiser 1987).

To estimate $\xi(\sigma, \pi)$ we must compare the observed count of galaxy pairs with the count estimated from a random distribution following the same selection function both on the

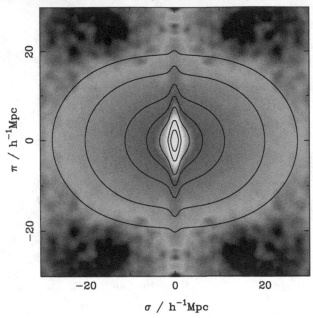

FIGURE 8. The galaxy correlation function $\xi(\sigma, \pi)$ as a function of transverse (σ) and radial (π) pair separation is shown as a greyscale image. It was computed in $0.2\,h^{-1}$ Mpc boxes and then smoothed with a Gaussian having an rms of $0.5\,h^{-1}$ Mpc. The contours are for a model with $\beta = 0.4$ and $\sigma_p = 400\,\mathrm{km\,s^{-1}}$, and are plotted at $\xi = 10, 5, 2, 1, 0.5, 0.2$ and 0.1.

sky and in redshift as the observed galaxies. Optimal weighting is applied to minimize the uncertainties due to cosmic variance and Poisson noise. This is close to equal-volume weighting out to an adopted redshift limit of $z = 0.25$. The results are robust against the uncertainties in both the survey mask and the weighting procedure. The redshift-space correlation function for the 2dFGRS computed in this way is shown in Figure 8. The correlation-function results display very clearly the two signatures of redshift-space distortions discussed above. The 'fingers of God' from small-scale random velocities are very clear, as indeed has been the case from the first redshift surveys (e.g. Davis & Peebles 1983). However, this is the first time that the large-scale flattening from coherent infall has been seen in detail.

The degree of large-scale flattening is determined by the total mass density parameter, Ω, and the biasing of the galaxy distribution. On large scales, it should be correct to assume a linear bias model, so that the redshift-space distortion on large scales depends on the combination $\beta \equiv \Omega^{0.6}/b$. In principle, this distortion should be a robust way to determine Ω (or at least β); see the reviews by Strauss & Willick (1995) and Hamilton (1997). On large scales, linear distortions should also be applicable, so we expect to see the following quadrupole-to-monopole ratio in the correlation function:

$$\frac{\xi_2}{\xi_0} = \frac{3+n}{n} \frac{4\beta/3 + 4\beta^2/7}{1 + 2\beta/3 + \beta^2/5} \; , \tag{6}$$

where n is the power spectrum index of the fluctuations, $\xi \propto r^{-(3+n)}$. This is modified by the Finger-of-God effect, which is significant even at large scales and dominant at small scales. The effect can be modeled by introducing a parameter σ_p, which represents the rms pairwise velocity dispersion of the galaxies in collapsed structures, σ_p (see e.g. Ballinger et al. 1996). For a survey that subtends a small angle (i.e. in the distant-

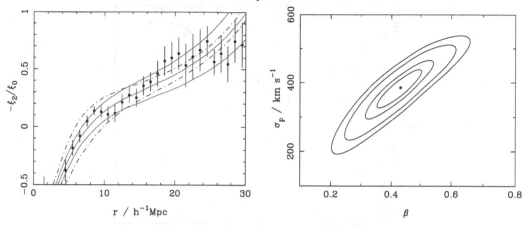

FIGURE 9. (a) The compression of $\xi(\sigma, \pi)$ as measured by its quadrupole-to-monopole ratio, plotted as $-\xi_2/\xi_0$. The solid lines correspond to models with $\sigma_p = 400\,\mathrm{km\,s^{-1}}$ and (bottom to top) $\beta = 0.3, 0.4, 0.5$, while the dot-dash lines correspond to models with $\beta = 0.4$ and (top to bottom) $\sigma_p = 300, 400, 500\,\mathrm{km\,s^{-1}}$. (b) Likelihood contours for β and σ_p from the model fits. The inner contour is the one-parameter 68% confidence ellipse; the outer contours are the two-parameter 68%, 95% and 99% confidence ellipses. The central dot is the maximum likelihood fit, with $\beta = 0.43$ and $\sigma_p = 385\,\mathrm{km\,s^{-1}}$.

observer approximation), a good approximation to the anisotropic redshift-space Fourier spectrum is given by multiplying the linear Kaiser anisotropy by a damping term from nonlinear effects:

$$\delta_k^s = \delta_k^r (1 + \beta \mu^2) D(k\sigma\mu) \ . \tag{7}$$

Here, $\beta = \Omega_m^{0.6}/b$, b being the linear bias parameter of the galaxies under study, and $\mu = \hat{\mathbf{k}} \cdot \hat{\mathbf{r}}$. For an exponential distribution of relative small-scale peculiar velocities (as seen empirically), the damping function is $D(y) \simeq (1 + y^2/2)^{-1/2}$, and $\sigma \simeq 400\,\mathrm{km\,s^{-1}}$ is a reasonable estimate for the pairwise velocity dispersion of galaxies. Full details of the procedure for fitting this expression to the 2dFGRS data are given in Peacock et al. (2001). The main assumption is that the form of the real-space power spectrum is as measured by the APM survey, although the results do not depend strongly on this choice.

Figure 9a shows the variation in ξ_2/ξ_0 as a function of scale. The ratio is positive on small scales where the Finger-of-God effect dominates, and negative on large scales where the Kaiser effect dominates. The best-fitting model (considering only the quasi-linear regime with $8 < r < 25\,h^{-1}\,\mathrm{Mpc}$) has $\beta \simeq 0.4$ and $\sigma_p \simeq 400\,\mathrm{km\,s^{-1}}$; the likelihood contours are shown in Figure 9b. Marginalizing over σ_p, the best estimate of β and its 68% confidence interval is

$$\beta = 0.43 \pm 0.07 \ . \tag{8}$$

This is the first precise measurement of β from redshift-space distortions; previous studies have shown the effect to exist (e.g. Hamilton, Tegmark & Padmanabhan 2000; Taylor et al. 2001; Outram, Hoyle & Shanks 2001), but achieved little more than 3σ detections.

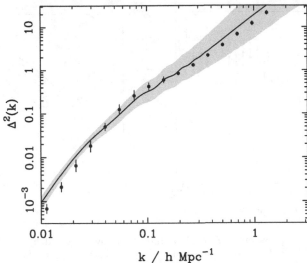

k / h Mpc^{-1}

FIGURE 10. The dimensionless matter power spectrum at zero redshift, $\Delta^2(k)$, as predicted from the allowed range of models that fit the microwave-background anisotropy data, plus the assumption that $H_0 = 70\,\mathrm{km\,s^{-1}Mpc^{-1}} \pm 10\%$. The solid line shows the best-fit model from Jaffe et al. (2001) [power-spectrum index $n = 1.01$, and density parameters in baryons, CDM, and vacuum of respectively 0.065, 0.285, 0.760]. The effects of nonlinear evolution have been included, according to a revised version of the procedure of Peacock & Dodds (1996). The shaded band shows the 1σ variation around this model allowed by the CMB data. The solid points are the real-space power spectrum measured for APM galaxies. The clear conclusion is that APM galaxies are consistent with being essentially unbiased tracers of the mass on large scales.

4. Cosmological parameters and the power spectrum

4.1. *Connection to CMB anisotropies*

The detailed measurement of the signature of gravitational collapse is the first major achievement of the 2dFGRS; we now consider the quantitative implications of this result. The first point to consider is that there may be significant corrections for luminosity effects. The optimal weighting means that the mean luminosity is high: it is approximately 1.9 times the characteristic luminosity, L^*, of the overall galaxy population (Folkes et al. 1999). Benoist et al. (1996) have suggested that the strength of galaxy clustering increases with luminosity, with an effective bias that can be fitted by $b/b^* = 0.7 + 0.3(L/L^*)$. This effect has been controversial (see Loveday et al. 1995), but the 2dFGRS dataset favors a very similar luminosity dependence. We therefore expect that β for L^* galaxies will exceed the directly measured figure. Applying a correction using the given formula for $b(L)$, we deduce $\beta(L = L^*) = 0.54 \pm 0.09$. Finally, the 2dFGRS has a median redshift of 0.11. With weighting, the mean redshift in the present analysis is $\bar{z} = 0.17$, and so the result should be interpreted as β at that epoch. The extrapolation to $z = 0$ is model-dependent, but probably does not introduce a significant change (Carlberg et al. 2000).

The 2dFGRS measurement of $\Omega^{0.6}/b$ would thus imply $\Omega = 0.36 \pm 0.10$ if L^* galaxies are unbiased, but it is difficult to justify such an assumption. In principle, the details of the clustering pattern in the nonlinear regime allow the $\Omega - b$ degeneracy to be broken (Verde et al. 1998), but for the present it is interesting to use an independent approach. Observations of CMB anisotropies can in principle measure almost all the cosmological parameters, and Jaffe et al. (2001) obtained the following values for the densities in collisionless matter (c), baryons (b), and vacuum (v): $\Omega_c + \Omega_b + \Omega_v = 1.11 \pm 0.07$, $\Omega_c h^2 =$

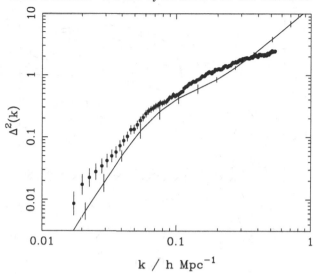

FIGURE 11. The 2dFGRS redshift-space power spectrum, estimated according to the FKP procedure. The solid points with error bars show the power estimate. The window function correlates the results at different k values, and also distorts the large-scale shape of the power spectrum An approximate correction for the latter effect has been applied. The line shows the real-space power spectrum estimated by deprojection from the APM survey.

0.14 ± 0.06, $\Omega_b h^2 = 0.032 \pm 0.005$, together with a power-spectrum index $n = 1.01 \pm 0.09$. The 2dFGRS result for β gives an independent test of this picture, as follows.

The only parameter left undetermined by the CMB data is the Hubble constant, h. Recent work (Mould et al. 2000; Freedman et al. 2001) indicates that this is now determined to an rms accuracy of 10%, and a central value of $h = 0.70$ is adopted hereafter. This completes the cosmological model, requiring a total matter density parameter $\Omega \equiv \Omega_c + \Omega_b = 0.35 \pm 0.14$. It is then possible to use the parameter limits from the CMB to predict a conservative range for the mass power spectrum at $z = 0$, which is shown in Figure 10. A remarkable feature of this plot is that the mass power spectrum appears to be in good agreement with the clustering observed in the APM survey (Baugh & Efstathiou 1994). For each model allowed by the CMB, we can predict both b (from the ratio of galaxy and mass spectra) and also β (since a given CMB model specifies Ω). Considering the allowed range of models, we then obtain the prediction $\beta_{\text{CMB+APM}} = 0.57 \pm 0.17$. A flux-limited survey such as the APM will have a mean luminosity close to L^*, so the appropriate comparison is with the 2dFGRS corrected figure of $\beta = 0.54 \pm 0.09$ for L^* galaxies. These numbers are in very close agreement. In the future, the value of β will become one of the most direct ways of confronting large-scale structure with CMB studies.

4.2. *The 2dFGRS power spectrum*

Of course, one may question the adoption of the APM power spectrum, which was deduced by deprojection of angular clustering. The 3D data of the 2dFGRS should be capable of improving on this determination, and a first attempt at doing this is shown in Figure 11. This power-spectrum estimate uses the FFT-based approach of Feldman, Kaiser & Peacock (1994), and needs to be interpreted with care. Firstly, it is a raw redshift-space estimate, so that the power beyond $k \simeq 0.2\, h\,\text{Mpc}^{-1}$ is severely damped by fingers of God. On large scales, the power is enhanced, both by the Kaiser effect and by

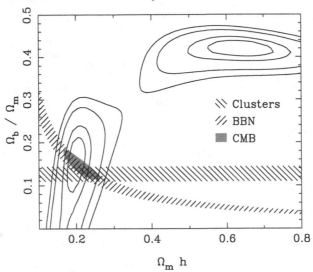

FIGURE 12. Likelihood contours for the best-fit linear power spectrum over the region $0.02 < k < 0.15$. The normalization is a free parameter to account for the unknown large scale biasing. Contours are plotted at the usual positions for one-parameter confidence of 68%, and two-parameter confidence of 68%, 95% and 99% (i.e. $-2\ln(\mathcal{L}/\mathcal{L}_{\max}) = 1, 2.3, 6.0, 9.2$). The results have been marginalized over the missing free parameters (h and the power spectrum amplitude) by integrating under the Likelihood surface. A prior on h of $h = 0.7 \pm 10\%$ was assumed. This result is compared to estimates from x-ray cluster analysis (Evrard 1997), big-bang nucleosynthesis (O'Meara et al. 2001) and recent CMB results. The CMB results assume that $\Omega_b h^2$ and $\Omega_{\mathrm{cdm}} h^2$ were independently determined from the data.

the luminosity-dependent clustering discussed above. Finally, the FKP estimator yields the true power convolved with the window function. This modifies the power significantly on large scales (roughly a 20% correction). An approximate correction for this has been made in Figure 11 by multiplying by the correction factor appropriate for a $\Gamma = 0.25$ CDM spectrum. The precision of the power measurement appears to be encouragingly high, and the systematic corrections from the window are well specified.

The next task is to perform a detailed fit of physical power spectra, taking full account of the window effects. The hope is that we will obtain not only a more precise measurement of the overall spectral shape, as parameterized by Γ, but will be able to move towards more detailed questions such as the existence of baryonic features in the matter spectrum (Meiksin, White & Peacock 1999). We summarize here results from the first attempt at this analysis (Percival et al. 2001).

The likelihood of each model has been estimated using a covariance matrix calculated from Gaussian realizations of linear density fields for a $\Omega_m h = 0.2$, $\Omega_b/\Omega_m = 0.15$ CDM power spectrum, for which $\chi^2_{\min} = 34.4$, given an expected value of 28. The best fit power spectrum parameters are only weakly dependent on this choice. The likelihood contours in Ω_b/Ω_m versus $\Omega_m h$ for this fit are shown in Figure 12. At each point in this surface, the results have been marginalized by integrating the Likelihood surface over the two free parameters, h and the power spectrum amplitude. The result is not significantly altered if instead, the modal, or Maximum Likelihood points in the plane corresponding to power spectrum amplitude and h were chosen. The likelihood function is also dependent on the covariance matrix (which should be allowed to vary with cosmology), although the consistency of result from covariance matrices calculated for different cosmologies shows that this dependence is negligibly small. Assuming a uniform prior for h over a factor of

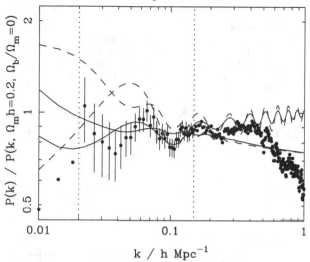

FIGURE 13. The 2dFGRS data compared with the two preferred models from the Maximum Likelihood fits convolved with the window function (solid lines). Error bars show the diagonal elements of the covariance matrix, for the fitted data that lie between the dotted vertical lines. The unconvolved models are also shown (dashed lines). The $\Omega_m h \simeq 0.6$, $\Omega_b/\Omega_m = 0.42$, $h = 0.7$ model has the higher bump at $k \simeq 0.05\,h\,\mathrm{Mpc}^{-1}$. The smoother $\Omega_m h \simeq 0.20$, $\Omega_b/\Omega_m = 0.15$, $h = 0.7$ model is a better fit to the data because of the overall shape.

2 is arguably over-cautious, and a Gaussian prior $h = 0.7 \pm 10\%$ may been considered. This corresponds to multiplying by the likelihood from external constraints such as the *HST* key project (Freedman et al. 2001); this has only a minor effect on the results.

Figure 12 shows that there is a degeneracy between $\Omega_m h$ and the baryonic fraction Ω_b/Ω_m. However, there are two local maxima in the likelihood, one with $\Omega_m h \simeq 0.2$ and $\sim 20\%$ baryons, plus a secondary solution $\Omega_m h \simeq 0.6$ and $\sim 40\%$ baryons. The high-density model can be rejected through a variety of arguments, and the preferred solution is

$$\Omega_m h = 0.20 \pm 0.03; \qquad \Omega_b/\Omega_m = 0.15 \pm 0.07 \ . \tag{9}$$

The 2dFGRS data are compared to the best-fit linear power spectra convolved with the window function in Figure 13. This shows where the two branches of solutions come from: the low-density model fits the overall shape of the spectrum with relatively small 'wiggles,' while the solution at $\Omega_m h \simeq 0.6$ provides a better fit to the bump at $k \simeq 0.065\,h\,\mathrm{Mpc}^{-1}$, but fits the overall shape less well.

Perhaps the main point to emphasize here is that the results are not greatly sensitive to the assumed tilt of the primordial spectrum. The CMB results motivate the choice of $n = 1$, but it is clear that very substantial tilts are required to alter the above conclusions significantly: $n \simeq 0.8$ would be required to turn zero baryons into the preferred model.

The conclusions certainly fit well with other constraints on the matter content of the universe, which are also shown in Figure 12. Latest estimates of the Deuterium to Hydrogen ratio in QSO spectra combined with big-bang nucleosynthesis theory predict $\Omega_b h^2 = 0.0205 \pm 0.0018$ (O'Meara et al. 2001), which matches the 2dFGRS result well, assuming $h = 0.7$ to convert to a baryon fraction as a function of Ωh. X-ray cluster analysis predicts a baryon fraction $\Omega_b/\Omega_m = 0.127 \pm 0.017$ (Evrard 1997) which is again within 1σ of the value implied by the 2dFGRS data. Figure 12 also shows the confidence region from CMB data, in this case averaging the results of Pryke et al. (2001) and

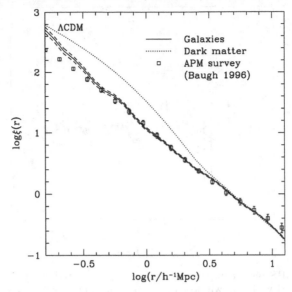

FIGURE 14. The correlation function of galaxies in the semianalytical simulation of an LCDM universe by Benson et al. (2000a). The predicted galaxy correlations are very close to a single power law, especially around $1\,h^{-1}\,\mathrm{Mpc}$, where the predicted mass correlations rise above the APM data.

Netterfield et al. (2001), which favor a slightly lower baryon density than Jaffe et al. (2001).

5. Galaxy formation and biased clustering

The disagreement between galaxy clustering and the mass clustering expected in CDM universes might be resolved if the relation between galaxies and the overall matter distribution is sufficiently complicated. Indeed, the formation of galaxies must be a non-local process to some extent, and the modern paradigm was introduced by White & Rees (1978): galaxies form through the cooling of baryonic material in virialized haloes of dark matter. The virial radii of these systems are in excess of 0.1 Mpc, so there is the potential for large differences in the correlation properties of galaxies and dark matter on these scales.

A number of studies have indicated that the observed galaxy correlations may indeed be reproduced by CDM models. The most direct approach is a numerical simulation that includes gas, and relevant dissipative processes. This is challenging, but just starting to be feasible with current computing power (Pearce et al. 1999). The alternative is 'semianalytic' modeling, in which the merging history of dark-matter haloes is treated via the extended Press-Schechter theory (Bond et al. 1991), and the location of galaxies within haloes is estimated using dynamical-friction arguments (e.g. Kauffmann et al. 1993, 1999; Cole et al. 1994; Somerville & Primack 1999; van Kampen, Jimenez & Peacock 1999; Benson et al. 2000a,b). Both these approaches have yielded similar conclusions, and shown how CDM models can match the galaxy data: specifically, the low-density flat ΛCDM model that is favored on other grounds can yield a correlation function that is close to a single power law over $1000 \gtrsim \xi \gtrsim 1$, even though the mass correlations show a marked curvature over this range (Pearce et al. 1999; Benson et al. 2000a; see Figure 14). These results are impressive, yet it is frustrating to have a result of such

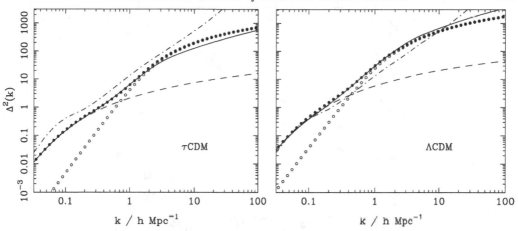

FIGURE 15. The power spectrum for the τCDM and ΛCDM models. The solid lines contrast the linear spectrum with the nonlinear spectrum, calculated according to the approximation of PD96. The spectrum according to randomly-placed haloes is denoted by open circles; if the linear power spectrum is added, the main features of the nonlinear spectrum are well reproduced. In neither case is the resulting nonlinear spectrum at all the same shape as the APM observations (shown as a dot-dashed line).

fundamental importance emerge from a complicated calculational apparatus. There is thus some motivation for constructing a simpler heuristic model that captures the main processes at work in the full semianalytic models. This section describes an approach of this sort (Peacock & Smith 2000; see also Seljak 2000 and Scoccimarro et al. 2001).

An early model for galaxy clustering was suggested by Neyman, Scott & Shane (1953), in which the nonlinear density field was taken to be a superposition of randomly-placed clumps. With our present knowledge about the evolution of CDM universes, this idealized model can be made considerably more realistic: hierarchical models are expected to contain a distribution of masses of clumps, which have density profiles that are more complicated than isothermal spheres. These issues are well studied in N-body simulations, and highly accurate fitting formulae exist, both for the mass function and for the density profiles. Briefly, the following calculations adopt the mass function of Sheth & Tormen (1999) and the halo profiles of Moore et al. (1999). Using this model, it is then possible to calculate the correlations of the nonlinear density field, neglecting only the large-scale correlations in halo positions. The power spectrum determined in this way is shown in Figure 15, and turns out to agree very well with the exact nonlinear result on small and intermediate scales. The lesson here is that a good deal of the nonlinear correlations of the dark matter field can be understood as a distribution of random clumps, provided these are given the correct distribution of masses and mass-dependent density profiles.

How can we extend this model to understand how the clustering of galaxies can differ from that of the mass? There are two distinct ways in which a degree of bias is inevitable:

(1)   Halo occupation numbers. For low-mass haloes, the probability of obtaining an L^* galaxy must fall to zero. For haloes with mass above this lower limit, the number of galaxies will in general not scale with halo mass.

(2)   Nonlocality. Galaxies can orbit within their host haloes, so the probability of forming a galaxy depends on the overall halo properties, not just the density at a point. Also, the galaxies will end up at special places within the haloes: for a

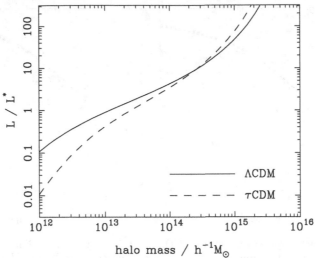

FIGURE 16. The empirical luminosity–mass relation required to reconcile the observed AGS luminosity function with two variants of CDM. L^* is the characteristic luminosity in the AGS luminosity function ($L^* = 7.6 \times 10^{10} h^{-2} L_\odot$). Note the rather flat slope around $M = 10^{13}$ to $10^{14} h^{-1} M_\odot$, especially for ΛCDM.

halo containing only one galaxy, the galaxy will clearly mark the halo center. In general, we expect one central galaxy and a number of satellites.

The numbers of galaxies that form in a halo of a given mass is the prime quantity that numerical models of galaxy formation aim to calculate. However, for a given assumed background cosmology, the answer may be determined empirically. Galaxy redshift surveys have been analyzed via grouping algorithms similar to the 'friends-of-friends' method widely employed to find virialized clumps in N-body simulations. With an appropriate correction for the survey limiting magnitude, the observed number of galaxies in a group can be converted to an estimate of the total stellar luminosity in a group. This allows a determination of the All Galaxy System (AGS) luminosity function: the distribution of virialized clumps of galaxies as a function of their total luminosity, from small systems like the Local Group to rich Abell clusters.

The AGS function for the CfA survey was investigated by Moore, Frenk & White (1993), who found that the result in blue light was well described by

$$d\phi = \phi^* \left[(L/L^*)^\beta + (L/L^*)^\gamma \right]^{-1} \, dL/L^* \ , \tag{10}$$

where $\phi^* = 0.00126 h^3 \mathrm{Mpc}^{-3}$, $\beta = 1.34$, $\gamma = 2.89$; the characteristic luminosity is $M^* = -21.42 + 5 \log_{10} h$ in Zwicky magnitudes, corresponding to $M_B^* = -21.71 + 5 \log_{10} h$, or $L^* = 7.6 \times 10^{10} h^{-2} L_\odot$, assuming $M_B^\odot = 5.48$. One notable feature of this function is that it is rather flat at low luminosities, in contrast to the mass function of dark-matter haloes (see Sheth & Tormen 1999). It is therefore clear that any fictitious galaxy catalogue generated by randomly sampling the mass is unlikely to be a good match to observation. The simplest cure for this deficiency is to assume that the stellar luminosity per virialized halo is a monotonic, but nonlinear, function of halo mass. The required luminosity–mass relation is then easily deduced by finding the luminosity at which the integrated AGS density $\Phi(> L)$ matches the integrated number density of haloes with mass $> M$. The result is shown in Figure 16.

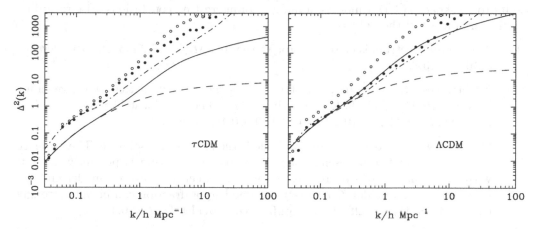

FIGURE 17. The power spectrum for galaxy catalogues constructed from the τCDM and ΛCDM models. A reasonable agreement with the APM data (dot-dashed line) is achieved by simple empirical adjustment of the occupation number of galaxies as a function of halo mass, plus a scheme for placing the haloes non-randomly within the haloes. In contrast, the galaxy power spectrum differs significantly in shape from that of the dark matter (linear and nonlinear theory shown as in Figure 15).

We can now return to the halo-based galaxy power spectrum and use the correct occupation number, N, as a function of mass. This is needs a little care at small numbers, however, since the number of haloes with occupation number unity affects the correlation properties strongly. These haloes contribute no correlated pairs, so they simply dilute the signal from the haloes with $N \geqslant 2$. The existence of antibias on intermediate scales can probably be traced to the fact that a large fraction of galaxy groups contain only one $> L_*$ galaxy. Finally, we need to put the galaxies in the correct location, as discussed above. If one galaxy always occupies the halo center, with others acting as satellites, the small-scale correlations automatically follow the slope of the halo density profile, which keeps them steep. The results of this exercise are shown in Figure 17, and are encouragingly similar to the scale-dependent bias found in the detailed calculations of Benson et al. (2000a), shown in Figure 14.

6. Conclusions

Although this talk has presented only a partial selection of the evidence, there is a good case for convergence in our understanding of the basis of large-scale structure. The power spectrum of the matter density that is predicted to result from gravitational instability is understood in detail, including small corrections from nonlinear effects and fine signatures of the baryon content. The great barrier to using large-scale structure as a tool for quantitative cosmology has long been the uncertainties associated with galaxy bias, and this problem is not yet beaten. Nevertheless, both the detailed semianalytic results and their incarnation in the generic 'halo model' give grounds for optimism that we may be starting to attain a physical understanding of the origin of galaxy bias. Assuming that this provisional understanding holds up, we should be able to use observations of the galaxy power spectrum with confidence in order to measure the properties of the matter density.

At present, the 2dFGRS data allow the galaxy power spectrum to be measured to high accuracy (10–15% rms) over about a decade in scale at $k < 0.15 \, h \, \mathrm{Mpc}^{-1}$. By fitting the results to the space of CDM models, a number of interesting conclusions arise regarding the matter content of the universe:

(1)   The power spectrum is close in shape to that of a $\Omega_m h = 0.2$ model, to a tolerance of about 20%.

(2)   Nevertheless, there is sufficient structure in the $P(k)$ data that the degeneracy between Ω_b/Ω_m and $\Omega_m h$ is weakly broken. The two local likelihood maxima have $(\Omega_m h, \Omega_b/\Omega_m) \simeq (0.2, 0.15)$ and $(0.6, 0.4)$ respectively.

(3)   Of these two solutions, the preferred one is the low-density solution. The evidence for detection of baryon oscillations in the power spectrum is presently modest, with a likelihood ratio of approximately 3 between the favored model and the best zero-baryon model. Conversely, a large baryon fraction can be very strongly excluded: $\Omega_b/\Omega_m < 0.28$ at 95% confidence, provided $\Omega_m h < 0.4$.

(4)   These conclusions do not depend strongly on the value of h, but they do depend on the tilt of the primordial spectrum, with $n \simeq 0.8$ being required to make a zero-baryon model the best fit.

(5)   The sensitivity to tilt emphasizes that the baryon signal comes in good part from the overall shape of the spectrum. Although the eye is struck by a single sharp 'spike' at $k \simeq 0.065 \, h \, \mathrm{Mpc}^{-1}$, the correlated nature of the errors in the $P(k)$ estimate means that such features tend not to be significant in isolation. The convolving effects of the window would require a very substantial spike in the true power in order to match the 2dFGRS data exactly. This is not possible within the compass of conventional models, and the conservative conclusion is that the apparent spike is probably enhanced by correlated noise. A proper statistical treatment is essential in such cases.

It is interesting to compare these conclusions with other constraints. According to Jaffe et al. (2001), the current CMB data require $\Omega_m h^2 = 0.17 \pm 0.06$, $\Omega_b h^2 = 0.032 \pm 0.005$, together with a power-spectrum index of $n = 1.01 \pm 0.09$, on the assumption of pure scalar fluctuations. If we take $h = 0.7 \pm 10\%$, this gives

$$\Omega_m h = 0.24 \pm 0.09; \qquad \Omega_b/\Omega_m = 0.19 \pm 0.07 \ , \tag{11}$$

in remarkably good agreement with the estimate from the 2dFGRS

$$\Omega_m h = 0.20 \pm 0.03; \qquad \Omega_b/\Omega_m = 0.15 \pm 0.07 \ . \tag{12}$$

It appears that we live in a universe that has $\Omega_m h \simeq 0.2$ with a baryon fraction of approximately 15%; it is hard to see how this conclusion can be seriously in error. Although the CDM model is claimed to have problems in matching galaxy-scale observations, it clearly works extremely well on large scales. Any new model that cures the small-scale problems will have to look very much like $\Omega_m = 0.3$ ΛCDM on large scales.

   This paper has included results from the 2dF Galaxy Redshift Survey. The members of the 2dFGRS team are Matthew Colless (ANU), John Peacock (ROE), Carlton M. Baugh (Durham), Joss Bland-Hawthorn (AAO), Terry Bridges (AAO), Russell Cannon (AAO), Shaun Cole (Durham), Chris Collins (LJMU), Warrick Couch (UNSW), Nicholas Cross (St. Andrews), Gavin Dalton (Oxford), Kathryn Deeley (UNSW), Roberto De Propris (UNSW), Simon Driver (St. Andrews), George Efstathiou (IoA), Richard S. Ellis (Caltech), Carlos S. Frenk (Durham), Karl Glazebrook (JHU), Carole Jackson (ANU), Ofer Lahav (IoA), Ian Lewis (AAO), Stuart Lumsden (Leeds), Steve Maddox (Nottingham),

Darren Madgwick (IoA), Peder Norberg (Durham), Will Percival (ROE), Bruce Peterson (ANU), Will Sutherland (ROE), Keith Taylor (Caltech). The survey was made possible by the dedicated efforts of the staff of the Anglo-Australian Observatory, both in creating the 2dF instrument, and in supporting it on the telescope.

REFERENCES

BALLINGER, W. E., PEACOCK, J. A., & HEAVENS, A. F. 1996 *MNRAS* **282**, 877.

BAUGH, C. M. & EFSTATHIOU, G. 1993 *MNRAS* **265**, 145.

BAUGH, C. M. & EFSTATHIOU. G. 1994 *MNRAS* **267**, 323.

BENOIST, C., MAUROGORDATO, S., DA COSTA, L. N., CAPPI, A., & SCHAEFFER, R. 1996 *ApJ* **472**, 452.

BENSON, A. J., COLE, S., FRENK, C. S., BAUGH, C. M., & LACEY, C. G. 2000a *MNRAS* **311**, 793.

BENSON, A. J., BAUGH, C. M., COLE, S., FRENK, C. S., & LACEY, C. G. 2000b *MNRAS* **316**, 107.

BOND, J. R., COLE, S., EFSTATHIOU, G., & KAISER, N. 1991 *ApJ* **379**, 440.

COLE, S., ARAGÓN-SALAMANCA, A., FRENK, C. S., NAVARRO, J. F., & ZEPF, S. E. 1994 *MNRAS* **271**, 781.

CARLBERG, R. G., YEE, H. K. C., MORRIS, S. L., LIN, H., HALL, P. B., PATTON, D., SAWICKI, M., & SHEPHERD, C. W. 2000 *ApJ* **542**, 57.

DAVIS, M. & PEEBLES, P. J. E. 1983 *ApJ* **267**, 465.

EFSTATHIOU, G., SUTHERLAND, W., & MADDOX, S. J. 1990 *Nature* **348**, 705.

EFSTATHIOU, G. & MOODY, S. 2001 *MNRAS* **325**, 1603.

EISENSTEIN, D. J. & HU, W. 1998 *ApJ* **496**, 605.

EISENSTEIN, D. J. & ZALDARRIAGA, M. 2001 *ApJ* **546**, 2.

EKE, V. R., COLE, S., & FRENK, C. S. 1996 *MNRAS* **282**, 263.

EVRARD, A. E. 1997 *MNRAS* **292**, 289.

FELDMAN, H. A., KAISER, N., & PEACOCK, J. A. 1994 *ApJ* **426**, 23.

FOLKES, S. J., ET AL. 1999 *MNRAS* **308**, 459.

FREEDMAN, W. L., ET AL. 2001 *ApJ* **553**, 47.

GAZTAÑAGA, E. & BAUGH, C. M. 1998 *MNRAS* **294**, 229.

GOLDBERG, D. M. & STRAUSS, M. A. 1998 *ApJ* **495**, 29.

HAMILTON, A. J. S. 1997 *astro-ph/9708102*.

HAMILTON, A. J. S., KUMAR, P., LU, E., & MATTHEWS, A. 1991 *ApL* **374**, 1.

HAMILTON, A. J. S., TEGMARK, M., & PADMANABHAN, N. 2000 *MNRAS* **317**, L23.

HOYLE, F., BAUGH, C. M., SHANKS, T., & RATCLIFFE, A. 1999 *MNRAS* **309**, 659.

JAFFE, A., ET AL. 2001 *Phys. Rev. Lett.* **86**, 3475.

JENKINS, A., ET AL. 1998 *ApJ* **499**, 20.

KAISER, N. 1987 *MNRAS* **227**, 1.

KAUFFMANN, G., WHITE, S. D. M., & GUIDERDONI, B. 1993 *MNRAS* **264**, 201.

KAUFFMANN, G., COLBERG, J. M., DIAFERIO, A., & WHITE, S. D. M. 1999 *MNRAS* **303**, 188.

KLYPIN, A., PRIMACK, J., & HOLTZMAN, J. 1996 *ApJ* **466**, 13.

LEWIS, I., TAYLOR, K., CANNON, R. D., GLAZEBROOK, K., BAILEY, J. A., FARRELL, T. J., LANKSHEAR, A., SHORTRIDGE, K., SMITH, G. A., GRAY, P. M., BARTON, J. R., MCCOWAGE, C., PARRY, I. R., STEVENSON, J., WALLER, L. G., WHITTARD, J. D., WILCOX, J. K., & WILLIS, K. C. 2001 *MNRAS*, submitted.

LOVEDAY, J., MADDOX, S. J., EFSTATHIOU, G., & PETERSON, B. A. 1995 *ApJ* **442**, 457.

MADDOX, S. J., EFSTATHIOU, G., SUTHERLAND, W. J., & LOVEDAY, J. 1990a, *MNRAS* **242**, 43.

MADDOX, S. J., SUTHERLAND, W. J., EFSTATHIOU, G., & LOVEDAY, J. 1990b, *MNRAS* **243**, 692.

MADDOX, S. J., EFSTATHIOU, G., & SUTHERLAND, W. J. 1990c *MNRAS* **246**, 433.

MADDOX, S., EFSTATHIOU, G., & SUTHERLAND, W. J. 1996 *MNRAS* **283**, 1227.

MEIKSIN, A. A., WHITE, M., & PEACOCK, J. A. 1999 *MNRAS* **304**, 851.

MOORE, B., FRENK, C. S., & WHITE, S. D. M. 1993 *MNRAS* **261**, 827.

MOORE, B., QUINN, T., GOVERNATO, F., STADEL, J., & LAKE, G. 1999 *MNRAS* **310**, 1147.

MOULD, J. R., ET AL. 2000 *ApJ* **529**, 786.

NEYMAN, J., SCOTT, E. L., & SHANE, C. D. 1953 *ApJ* **117**, 92.

NETTERFIELD, C. B., ET AL. 2001 *astro-ph/0104460*.

O'MEARA, J. M., ET AL. 2001 *ApJ* **552**, 718.

OUTRAM, P. J., HOYLE, F., & SHANKS, T. 2001, *MNRAS* **321**, 497.

PEACOCK, J. A. & DODDS, S. J. 1994 *MNRAS* **267**, 1020.

PEACOCK, J. A. & DODDS, S. J. 1996 *MNRAS* **280**, L19.

PEACOCK, J. A. & SMITH, R. E. 2000 *MNRAS* **318**, 1144.

PEACOCK, J. A. 1997 *MNRAS* **284**, 885.

PEACOCK, J. A., ET AL. 2001 *Nature* **410**, 169.

PEARCE, F. R., ET AL. 1999 *ApJ* **521**, L99.

PERCIVAL, W. J., ET AL. 2001 *MNRAS* **327**, 1297.

PIERPAOLI, E., SCOTT, D., & WHITE, M. 2001 *MNRAS* **325**, 77.

PRYKE, C., ET AL. 2001 *astro-ph/0104490*.

REIPRICH, T. H. & BÖHRINGER, H. 2001 *astro-ph/0111285*.

SAUNDERS, W., ET AL. 2000 *MNRAS* **317**, 55.

SCHLEGEL, D. J., FINKBEINER, D. P., & DAVIS, M. 1998 *ApJ* **500** 525.

SELJAK, U. 2000 *MNRAS* **318**, 203.

SELJAK, U. 2001 *astro-ph/0111362*.

SCOCCIMARRO, R., SHETH, R. K., HUI, L., & JAIN, B. 2001 *ApJ* **546**, 20.

SHECTMAN, S. A., LANDY, S. D., OEMLER, A., TUCKER, D. L., LIN, H., KIRSHNER, R. P., & SCHECHTER, P. L. 1996 *ApJ* **470**, 172.

SHETH, R. K. & TORMEN, G. 1999 *MNRAS* **308**, 119.

SOMERVILLE, R. S. & PRIMACK, J. R. 1999 *MNRAS* **310**, 1087.

STRAUSS, M. A. & WILLICK, J. A. 1995 *Physics Reports* **261**, 271.

TAYLOR, A. N., BALLINGER, W. E., HEAVENS, A. F., & TADROS, H. 2001 *MNRAS* **327**, 689.

VAN KAMPEN, E., JIMENEZ, R., & PEACOCK, J. A. 1999 *MNRAS* **310**, 43.

VERDE, L., HEAVENS, A. F., MATARRESE, S., MOSCARDINI, L. 1998 *MNRAS* **300**, 747.

VIANA, P. T. & LIDDLE, A. R. 1996 *MNRAS* **281**, 323.

VIANA, P. T., NICHOL, R. C., & LIDDLE, A. R. 2001 *astro-ph/0111394*.

WHITE, S. D. M. & REES, M. 1978 *MNRAS* **183**, 341.

WHITE, S. D. M., EFSTATHIOU, G., & FRENK, C. S. 1993 *MNRAS* **262**, 1023.

Cosmological implications of the most distant supernova (known)

By ADAM G. RIESS

Space Telescope Science Institute, 3700 San Martin Drive, Baltimore, MD 21218, USA

We present photometric observations of an apparent Type Ia supernova (SN Ia) at a redshift of ~1.7, the farthest SN observed to date. The supernova, SN1997ff, was discovered in a repeat observation by the *Hubble Space Telescope* (*HST*) of the Hubble Deep Field-North (HDF-N), and serendipitously monitored with NICMOS on *HST* throughout the Thompson et al. GTO campaign. The SN type can be determined from the host galaxy type: an evolved, red elliptical lacking enough recent star formation to provide a significant population of core-collapse supernovae. The classification is further supported by diagnostics available from the observed colors and temporal behavior of the SN, both of which match a typical SN Ia. The photometric record of the SN includes a dozen flux measurements in the I, J, and H bands spanning 35 days in the observed frame. The redshift derived from the SN photometry, $z = 1.7 \pm 0.1$, is in excellent agreement with the redshift estimate of $z = 1.65 \pm 0.15$ derived from the $U_{300}B_{450}V_{606}I_{814}J_{110}J_{125}H_{160}H_{165}K_s$ photometry of the galaxy. Optical and near-infrared spectra of the host provide a very tentative spectroscopic redshift of 1.755. Fits to observations of the SN provide constraints for the redshift-distance relation of SNe Ia and a powerful test of the current accelerating Universe hypothesis. The apparent SN brightness is consistent with that expected in the decelerating phase of the preferred cosmological model, $\Omega_M \approx 1/3, \Omega_\Lambda \approx 2/3$. It is inconsistent with grey dust or simple luminosity evolution, candidate astrophysical effects which could mimic previous evidence for an accelerating Universe from SNe Ia at $z \approx 0.5$. We consider several sources of potential systematic error including gravitational lensing, supernova misclassification, sample selection bias, and luminosity calibration errors. Currently, none of these effects alone appears likely to challenge our conclusions. Additional SNe Ia at $z > 1$ will be required to test more exotic alternatives to the accelerating Universe hypothesis and to probe the nature of dark energy.

1. Introduction

In the past five years the *Hubble Space Telescope* (*HST*) has played a central role in the study of Type Ia supernovae (SNe Ia) at $0.4 < z < 1.0$. Observations by the High-z Team (HZT; Schmidt et al. 1998) and those by the Supernova Cosmology Project (SCP; Perlmutter et al. 1998) indicate that the Universe is not decelerating as expected for an $\Omega_M = 1$ universe (Garnavich et al. 1998; Perlmutter et al. 1998). Far more startling and puzzling is the evidence that the Universe now appears to be *accelerating* (Riess et al. 1998; Perlmutter et al. 1999), propelled by a vacuum energy with negative pressure. The direct evidence for cosmic acceleration comes from the apparent faintness of SNe Ia at $z \approx 0.5$. Measurements of the characteristic angular scale of fluctuations in the cosmic microwave background (CMB) show $\Omega_{\text{total}} = 1.00 \pm 0.04$, which requires $\Omega_\Lambda > 0$ or something much like it, if $\Omega_M \approx 0.3 \pm 0.1$ (de Bernardis et al. 2000; Balbi et al. 2000). If cosmic acceleration is real, the physics community may well need to look beyond the Standard Model. Candidates for the dark energy include Einstein's cosmological constant, evolving scalar fields (modern cousins of the inflation field; Caldwell, Dave, & Steinhardt 1998), and a weakening of gravity in our $3 + 1$ dimensions due to the higher dimensions required by string theory (Deffayet, Dvali, & Gabadadze 2001). These explanations bear so greatly on fundamental physics that stronger evidence for dark energy is essential.

Contaminating astrophysical effects can imitate the evidence for an accelerating Universe. A pervasive screen of grey dust could dim SNe Ia with little telltale reddening apparent from their observed colors (Aguirre 1999a,b; Rana 1979, 1980). Although the first exploration of a distant SN Ia at near-infrared (near-IR) wavelengths provided no evidence of nearly grey dust, more data are needed to perform a definitive test (Riess et al. 2000).

A more familiar challenge to the measurement of the global acceleration or deceleration rate is luminosity evolution (Sandage & Hardy 1973). The lack of a complete theoretical understanding of SNe Ia and an inability to identify their specific progenitor systems undermines our ability to predict with confidence the direction or degree of luminosity evolution (Höflich, Wheeler, & Thielemann 1998; Umeda et al. 1999a,b; Livio 2000; Drell, Loredo, & Wasserman 2000; Pinto & Eastman 2000; Yungelson & Livio 2000). The weight of empirical evidence appears to disfavor evolution as an alternative to dark energy, as the cause of the apparent faintness of SNe Ia at $z \approx 0.5$ (see Riess 2000 for a review). However, the case against evolution remains short of compelling.

The extraordinary claim of the existence of dark energy requires a high level of evidence for its acceptance. Fortunately, a direct and definitive test is available. It should be possible to discriminate between cosmological models and "impostors" by tracing the redshift-distance relation to redshifts greater than one.

1.1. *The Next Redshift Octave and the Epoch of Deceleration*

If the cosmological acceleration inferred from SNe Ia is real, it commenced rather recently, at $0.5 < z < 1$. Beyond these redshifts, the Universe was more compact and the attraction of matter dominated the repulsion of dark energy. At $z > 1$ the expansion of the Universe should have been decelerating (see Filippenko & Riess 2000). The observable result at $z \geq 1$ would be an apparent *increased* brightness of SNe Ia relative to what is expected for a non-decelerating Universe. However, if the apparent faintness of SNe Ia at $z \approx 0.5$ is caused by dust or simple evolution, SNe Ia at $z > 1$ should appear fainter than expected from decelerating cosmological models. More complex parameterizations of evolution or extinction which can match both the accelerating and decelerating epochs of expansion would require a higher order of fine tuning and are therefore less plausible.

Both the Supernova Cosmology Project (SCP; Perlmutter et al. 1995) and the High-z Supernova Search Team (HZT; Schmidt et al. 1998) have pursued the discovery of SNe Ia in this next redshift interval. In the fall of 1998 the SCP reported the discovery of SN1998ef at $z = 1.2$ (Aldering et al. 1998). The following year the HZT discovered a SN Ia at $z = 1.2$ (SN1999fv) as well as at least one more at $z \approx 1.05$ (Tonry et al. 1999; Coil et al. 2000). These data sets, while currently lacking the statistical power to discriminate between cosmological and astrophysical effects, are growing, and in the future may provide the means to break degeneracies.

In early 1998, Gilliland & Phillips (1998) reported the detection of two SNe, SN1997ff and SN1997fg, in a reobservation of the Hubble Deep Field-North (HDF-N) with WFPC2 through the F814W filter. The elliptical host of SN1997ff indicated that this supernova was "most probably a SN Ia ...[at] the greatest distance reported previously for SNe," but the observations at a single epoch and in a single band were insufficient to provide useful constraints on the SN and hence to perform cosmological tests (Gilliland, Nugent, & Phillips 1999, hereafter GNP99).

At this conference we report additional, *serendipitous* observations of SN1997ff obtained in the Guaranteed Time Observer (GTO) NICMOS campaign (Thompson et al. 1999) and in General Observer (GO) program 7817 (Dickinson et al. 2001), as well as spectroscopy of the host. The combined data set provides the ability to put strong constraints on the redshift and distance of this supernova and shows it to be the highest-redshift SN Ia observed (to date). These measurements further provide an opportunity to perform a new and powerful test of the accelerating Universe by probing its preceding epoch of deceleration.

2. SN1997ff

Two near-IR assaults on the HDF-N with NICMOS on *HST* provided a wealth of data and understanding on the natural history of galaxies (see Ferguson, Dickinson, & Williams 2000 for a review). The GTO program of Thompson et al. (1999; GTO 7235) consisted of ~100 orbits of F110W and F160W exposures of a single $55'' \times 55''$ Camera-3 field, reaching a limiting AB magnitude (Oke & Gunn 1983) of 29 in the latter. The observations were gathered during 14 consecutive days and the field was contained within the WF4 portion of HDF-N, *serendipitously* imaging the host of SN1997ff. (It is interesting to note that the placement of the GTO field within the HDF-N had less than a 20% chance of containing SN1997ff.) Although the program did not begin until January 19, 1998, about 25 days after the discovery of the SN, a series of single-dither exposures (GO 7807) was taken between the discovery of the SN and the start of the GTO program for the purpose of verifying the suitability of the chosen guide stars. Each of these exposures was for a duration 960 s. A single F110W and F240M exposure on Jan. 6, 1998 included the host as did a F160W exposure from Jan. 2, 1998 and another on Dec. 26, 1997. *The Dec. 26 NICMOS exposure was coincident within hours of the WFPC2 discovery exposures.* (It is of further interest to note the low likelihood of the chance temporal coincidence of the HDF-N SN Search and the GTO program, each initially scheduled in different *HST* cycles.)

A second program was undertaken six months after the GTO program, between June 14 and June 22, 1998 by Dickinson et al. (2001; GO 7817). This program observed the entire HDF-N in F110W and F160W to a limiting AB magnitude of ~26.5 by mosaicing Camera 3 of NICMOS to study a wider field of galaxies. This program also contained the host galaxy and the greatly faded light of the supernova. The space-based near-IR photometry of the SN host offered greater precision and coverage of the SED than the ground-based data alone and allowed an improved estimate of the photometric redshift. Using the space-based $U_{300}B_{450}V_{606}I_{814}J_{110}H_{160}$ photometry and the ground-based $J_{125}H_{165}K_s$ photometry contained in Table 1, Budavári et al. (2000) determined the redshift of the host to be $z = 1.65 \pm 0.15$ from fits to either galaxy SED eigenspectra or these same eigenspectra mildly corrected to improve the agreement between spectroscopic and photometric redshifts.

The exposure time for the SN in the Thompson et al. campaign was dispersed irregularly over a 35-day time interval. Although the now standard method of galaxy subtraction was used to measure the photometry of the SN, the application of this method was far more difficult than usual and required a number of special considerations discussed thoroughly in Riess et al. (2001). We direct those interested in the data reduction details to that work. Images of the SN and region of the host can be seen in Figure 1.

We measured the flux of the SN across the GTO campaign. For the NICMOS observations of SN1997ff, the dominant source of uncertainty is the host galaxy residuals in the difference image. Flux uncertainties were determined by a Monte Carlo exercise of adding and measuring artificial SNe in the field with the same brightness and background as SN1997ff (Schmidt et al. 1998). We transformed the relative SN flux onto the F110W AB and F160W AB magnitude systems by applying the zero points of the transformation equation from Dickinson et al. (2001). The Vega-system F160W and F110W magnitudes are given in Riess et al. (2001).

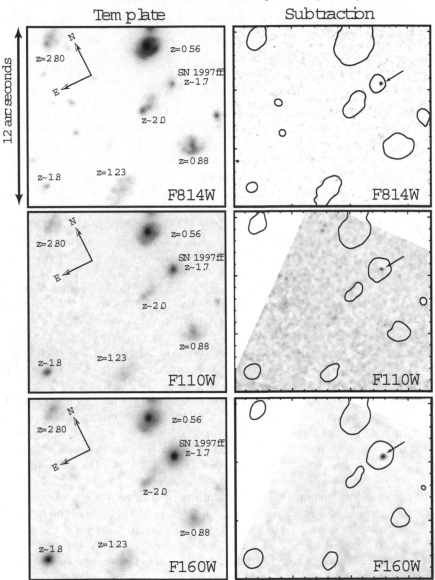

FIGURE 1.

Optical spectroscopy were obtained with ESI and LRIS on *Keck*. The total data set of optical spectroscopy consisted of 8 hours with a wavelength coverage of 4000 Å to 10000 Å. A small amount of continuum flux was detected in the composite spectrum with evidence of some minor breaks in the galaxy SED. Observations were made with the near-IR spectrograph NIRSPEC (McLean et al. 1998) on *Keck II* on the nights of 16 and 17 March 2001 UT and on 14 April 2001 UT. The feature which may be *very* tentatively identified as the [O II] $\lambda 3727$ line at $z = 1.755$ falls in a relatively rare region that is unaffected by bright OH lines in the night sky, although there is a strong sky emission feature at 1.029 μm, i.e. just to the red of the putative [O II] feature. This feature is noteworthy and may function as a useful hypothesis to test with future observations, but at this time its validity is highly uncertain.

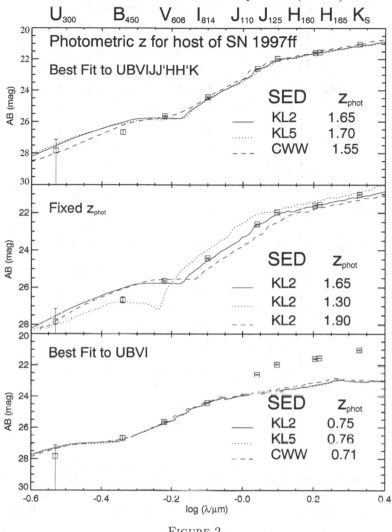

FIGURE 2.

3. Analysis

Fitting the $U_{300}B_{450}V_{606}I_{814}J_{110}H_{160}$ space-based photometry and the $J_{125}H_{165}K_s$ ground-based photometry of 4-403.0 to galaxy SEDs yields a photometric redshift of $z = 1.55$ to 1.70 with variations depending on whether the fitted model is based on template galaxy SEDs (Coleman, Wu, & Weedman 1980, hereafter CWW) or galaxy eigenspectra (Budavári et al. 2000). These fits can be seen in Figure 2.

An additional and independent pathway to determine the redshift is from the SN colors. For the following analysis we will provisionally adopt the classification from GNP99 of SN1997ff as a Type Ia supernova based on its red, elliptical host galaxy. However, later we will analyze the degree to which this classification is merited.

Coincident or near-coincident measurements of SN1997ff in different bands provide an observed $I - H$ color of 3.5 ± 0.2 mag and a $J - H$ color of 1.6 ± 0.2 mag, $30/(1 + z)$ days later in the rest frame. In Figure 3 we plot these measurements as a function of expected colors of SNe Ia over their temporal evolution at different redshifts (note that the size of the point scales with the temporal proximity to B maximum). SNe Ia are bluest shortly

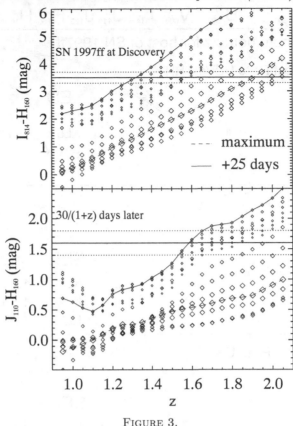

FIGURE 3.

after explosion and become redder with age. They reach their reddest color ∼25 days after maximum in these bandpasses and return to a modestly bluer color during the subsequent nebular phase. As seen in Figure 3, either of the observed colors of SN1997ff is redder than a SN Ia at any phase for $z < 1$. The $I - H$ color sets a limit of $z > 1.2$ while the $J - H$ color is more stringent with $z > 1.4$, both at the >95% confidence level.

This simple analysis assumes negligible reddening from the host. While this assumption is appropriate for most elliptical hosts, we will consider the effect of reddening explicitly in §4.2. Note that Galactic reddening is very low toward the HDF-N.

The above exercise cannot readily be performed if we assume the SED of a common SN II instead of a SN Ia. Not surprisingly, a comparison of the observed colors of SN1997ff to those expected for a blue SN II (similar to the well-observed SN1979C; e.g. Schmidt et al. 1994) yields poor fits to both color measurements and their time separation. For any value of the redshift, the observed $J - H$ color is far bluer than the color of a common SN II-P or SN II-L (for definitions, see Barbon, Ciatti, & Rosino 1979) at a phase dictated by the requirement of matching the earlier $I - H$ color. A common SN II would not match both color measurements of SN1997ff unless it were observed shortly after explosion and at $z \approx 2$. However, the observed decline of SN1997ff appears inconsistent with the expected rising luminosity or plateau phase of a SN II shortly after explosion. Reddening of the SN would result in an even greater difference between the data and a common SN II. Based on the observed colors and declining luminosity alone, the identification of SN1997ff as a normal SN II is strongly disfavored, a conclusion which is further discussed in §4.2.

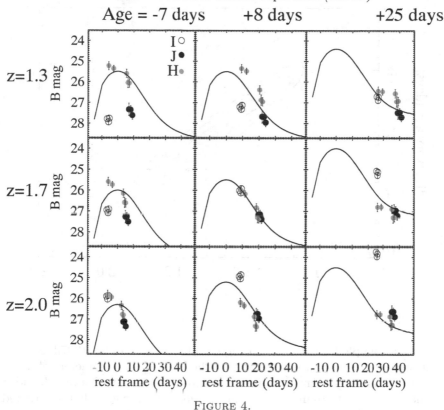

FIGURE 4.

The spectroscopy of the host presented provides some evidence which is consistent with the preceding redshift determinations and is potentially more precise, but is currently unreliable. The optical spectrum of the host suggests two minor breaks which could be identified with the rest-frame breaks at 2640 Å and 2900 Å due to blends of metals (Spinrad et al. 1997). A simple χ^2 minimization between the optical spectrum of the host and the same region of the SED of the $z = 1.55$ elliptical LBDS 53w091 (Spinrad et al. 1997) yields a significant minimum bounded by $z = 1.67$ and 1.79 (3σ confidence level). However, this minimum does not appear robust and we cannot rule out the possibility that other redshift matches are possible given other models for the host SED.

Although the spectroscopic redshift indicators are suggestive of a match with the photometric indicators, the quality of the spectra is too low and the identification of spectral features too uncertain to reach a robust determination of the redshift from the spectroscopy alone. Therefore in the following we derive constraints from the SN without employing the spectroscopic redshift indications.

The simple method for constraining the redshift described previously can be refined to make use of all of the SN photometric data simultaneously. By varying the parameters needed to empirically fit a SN Ia, such as the light-curve shape, distance, redshift, and age, we can use the quality of the fit to determine the probability density function (PDF) of these parameters. An additional component of this fitting process is to include the known correlation between SN Ia light-curve shapes and their peak luminosities (Phillips 1993; Phillips et al. 1999; Riess, Press, & Kirshner 1996; Perlmutter et al. 1997). Examples of this fitting process can be seen in Figure 4 as applied to SN1997ff.

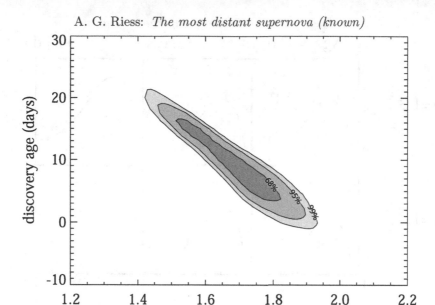

FIGURE 5.

We have used a simple formalism for using the observations of SN 1997ff and any prior information which is appropriate to determine the PDF of the parameters of luminosity, distance, redshift, and age commonly used to empirically model SNe Ia. The advantage of this method is its ability to incorporate prior information (e.g. a photometric redshift) in a statistically sound way which properly assigns weights to the relative constraints provided by data and priors.

Due to the presence of SN light in the Dickinson et al. (2001) template images taken 177 days after the SN discovery, it is necessary to restore the flux which is necessarily oversubtracted from the F110W and F160W magnitudes. The size of the correction depends on the redshift, age of discovery, and shape of the SN light curve. Therefore, we implemented this correction during the process of fitting the data to parameterized models. We determined the PDF for SN1997ff using our simple Bayesian formalism.

The marginalized PDF of the redshift is not a simple function, though it is strongly peaked near $z \approx 1.7$ and is insignificant outside the range $1.4 < z < 1.95$. A much lower local maximum is seen at $z = 1.55$. The redshift measurement of SN1997ff can be crudely approximated by $z = 1.7^{+0.10}_{-0.15}$.

The consistency of the three redshift indicators, determined independently from the galaxy colors, the supernova colors, and the host spectroscopy, provides a powerful and successful crosscheck of our redshift determination. Excluding any galaxy redshift information has little impact on the marginalized redshift PDF of the SN because the SN data are significantly more constraining for the redshift determination. The cause of the difference in measured photometric redshift precision lies in the difference in the relative homogeneity of galaxy and SN Ia colors. For the galaxy photometric determination, the precision of this method is limited by the variations of galaxy SEDs beyond those which can even be accounted for from the superposition of eigenspectra. In contrast, SNe Ia colors are far more homogeneous and their mild inhomogeneities are well characterized, leading to more precise constraints on the photometric redshift.

From the redshift determination we conclude that SN1997ff is the highest-redshift SN observed to date (as suspected by GNP99), easily surpassing the two SNe Ia at $z = 1.2$

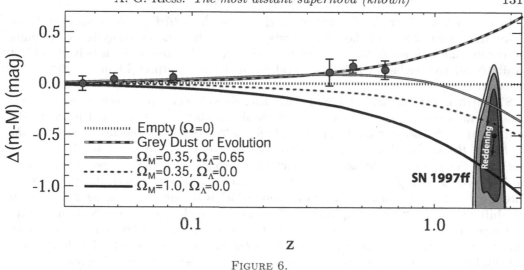

FIGURE 6.

from the HZT (Tonry et al. 1999; Coil et al. 2000) and the SCP (Aldering et al. 1998). The statistical confidence in this statement is very high.

We conclude that the SN was discovered by Gilliland & Phillips (1998) at an age of a week past B-band maximum with an uncertainty of ~5 days. An additional, local maximum in the marginalized probability is evident at an age of ~15 days after maximum. The possibility that the SN was discovered at this later age corresponds to the same model for which the weaker maximum in the redshift PDF indicated that $z = 1.55$. This correlation between the redshift and age parameters is a natural consequence of the reddening of a SN Ia as it ages, and is shown in Figure 5.

Of critical importance to cosmological hypothesis testing is the determination of the luminosity distance to such high-redshift SNe Ia. Because of the significant correlation between the distance and photometric redshift parameters, and the need to use both of these parameters for cosmological applications, we determined the 2-D PDF for distance and redshift, simultaneously. We also determined the likelihood function for the distance assuming the tentative spectroscopic redshift. We find $m - M = 45.15 \pm 0.34$ mag. This likelihood function is quite Gaussian within the 2σ boundaries, but flattens beyond for shorter distances (corresponding to an older discovery age) and steepens beyond for longer distances (corresponding to a discovery near maximum).

In Figure 6 we show the redshift and distance data (i.e. the Hubble diagram) for SNe Ia as presented by the Supernova Cosmology Project (Perlmutter et al. 1999) and the High-z Supernova Search Team (Riess et al. 1998). These data have been binned in redshift to depict the statistical leverage of the SN Ia sample. Overplotted are the cosmological models $\Omega_M = 0.35$, $\Omega_\Lambda = 0.65$ (favored), $\Omega_M = 0.35$, $\Omega_\Lambda = 0.0$ (open), $\Omega_M = 1.00$, $\Omega_\Lambda = 0.0$ (Einstein-de Sitter), and an astrophysical model representing a progressive dimming in proportion to redshift due to grey dust or simple evolution within an open cosmology. This model is further described in §4.2. All data and models are plotted as their difference from an empty Universe ($\Omega_M = 0.0$, $\Omega_\Lambda = 0.0$).

All models are equivalent in the limit of $z = 0$. Differences in the models are considerable and detectable at $z > 0.1$. Evidence for a significant dark energy density and current acceleration is provided by the excessive faintness of the binned data with $0.3 < z < 0.8$ compared to the open model, a net difference of ~0.25 mag.

In Figure 6 we show the constraints derived from SN1997ff. In the redshift-distance plane the principal axes of the error matrix from the photometric analysis are not quite perpendicular and the confidence contours are complex. Because there is only one object available in this highest redshift interval we prefer to interpret Figure 6 with broad brushstrokes.

SN1997ff is brighter by $\sim$1.1 mag (and therefore closer) than expected for the persistence of a purported source of astrophysical dimming at $z \approx 0.5$ and beyond. The statistical confidence of this statement is high ($> 99.99\%$). *This conclusion supports the reality of the measured acceleration of the Universe from SNe Ia at $z \approx 0.5$ by excluding the most likely, simple alternatives.* To avoid this conclusion requires the addition of an added layer of astrophysical complexity (e.g. intergalactic dust which dissipates in the interval $0.5 < z < 1.7$, or luminosity evolution which is suppressed or changes sign in this redshift interval). Other astrophysical effects such as a change in the SN Ia luminosity distance due to a change in metallicity with redshift are also disfavored (Shanks et al. 2001). Systematic challenges to these conclusions are addressed in §4. Models with relatively high vacuum energy and relatively low mass density are excluded (e.g. $\Omega_\Lambda \approx 1$, $\Omega_M \approx 0$). If we assume an approximately flat cosmology as required by observations of the CMB, and a cosmological constant-like nature for dark energy, the observations of SN1997ff disfavor $\Omega_\Lambda > 0.85$ or alternately $\Omega_M < 0.15$.

SN1997ff also provides an indication that the Universe was decelerating at the time of the supernova's explosion. To better understand this likelihood, we consider the redshift-distance relation of SNe Ia compared to a family of flat, Ω_Λ cosmologies. For such cosmologies, the transition redshift between the accelerating and decelerating epochs occurs at a redshift of $[2\Omega_\Lambda/\Omega_M]^{1/3} - 1$ (M. Turner, 2001, private communication). For increasing values of Ω_Λ, the transition point (i.e. the coasting point) occurs at increasing redshifts. The highest value of Ω_Λ which is marginally consistent with SN1997ff is $\Omega_\Lambda = 0.85$ (at the $\sim$3σ confidence level), for which the transition redshift occurs at $z = 1.25$, significantly below the redshift of SN1997ff. For the Universe to have commenced accelerating before the explosion of SN1997ff requires a value of $\Omega_\Lambda > 0.9$, a result which is highly in conflict with the SN brightness. We conclude that, within the framework of these simple but plausible cosmological models, SN1997ff exploded when the Universe was still decelerating. Indeed, the increase in the measured luminosity distance of SNe Ia between $z \approx 0.5$ and $z \approx 1.7$, a factor of 4.0, is significantly smaller than in most eternally coasting cosmologies (e.g. $\Omega_M = 0$, $\Omega_\Lambda = 0$) and appears to favor the empirical reality of a net deceleration over this range in redshift. However, a rigorous and quantitative test of past deceleration requires a more complete consideration of the possible nature of dark energy and is beyond the scope of this paper.

The above conclusions are unchanged if we adopt the tentative spectroscopic redshift of the SN host in place of the photometric redshift indicators. In this case the redshift uncertainty is greatly diminished and the distance uncertainty is mildly reduced. However, the dominant source of statistical uncertainty in the testing of cosmological hypothesis remains the distance uncertainty, not the redshift uncertainty.

4. Discussion

The results indicate that SN1997ff is the most distant SN Ia observed to date with a redshift of $z = 1.7^{+0.1}_{-0.15}$. Moreover, an estimate of its luminosity distance is consistent with an earlier epoch of deceleration and is inconsistent with astrophysical challenges (e.g. simple evolution or grey dust) to the inference of a currently accelerating Universe from SNe Ia at $z \approx 0.5$. Now we explore systematic uncertainties in these conclusions.

4.1. *SN Classification*

An alternate way to discriminate some SNe Ia is from the morphology of their host galaxy and its associated star-formation history. While all types of SNe have been observed in late-type galaxies, *SNe Ia are the only type to have been observed in early-type galaxies*. Although this lore is well known by experienced observers of SNe, this correlation is empirically apparent from an update of the Asiago SN catalog (Cappellaro et al. 1997; Asiago web site). Of the >1000 SNe whose type and host galaxy morphologies are all well-defined and have been classified in the modern scheme, there have been no core-collapse SNe observed in early-type galaxies (that is, only SNe Ia have been found in such galaxies). All ~40 SNe in elliptical hosts, and classified since the identification of the SN Ia sub-type, have been SNe Ia. The same homogeneity of type is true for the ~ 40 SNe classified in S0 hosts. Core-collapse SNe (types II, Ib, and Ic) first appear along the Hubble sequence in Sa galaxies, and even within these hosts they form a minority and their relative frequency to SNe Ia is suppressed by a factor of ~6 compared to their presence in late-type spirals (Cappellaro et al. 1997, 1999). Evolved systems lose their ability to produce core-collapse SNe.

The explanation for this well-known observation is deeply rooted in the nature of supernova progenitors and their ages. Unlike all other types of SNe which result from core collapse in massive stars, SNe Ia are believed to arise from the thermonuclear disruption of a white dwarf near the Chandrasekhar limit and thus occur in evolved stellar populations (see Livio 2000 for a review). The loss of massive stars in elliptical and S0 galaxies, without comparable replacement, quenches the production of core-collapse SNe, while SNe Ia, arising from relatively old progenitors, persist. From the host type of SN1997ff we might readily conclude, as did GNP99, that it is of Type Ia.

However, more careful consideration is needed to classify SN1997ff. Due to its high red-shift, we need to determine the degree of ongoing star formation and hence the likelihood of the appearance of a core-collapse SN from a young, massive star.

We analyzed the SFH by comparing the complete ultraviolet-optical-infrared SED of the host galaxy to Bruzual & Charlot (1993) population synthesis models. The upper panel of Figure 7 superimposes the galaxy photometry with models that assume a single, short burst of star formation with a Salpeter initial mass function. Such a model has negligible ongoing star formation after the initial burst, and thus its rest-frame ultraviolet (UV) to optical colors redden as quickly as possible. After 1 Gyr has elapsed, this model approximately matches the observed-frame colors of the host galaxy; a burst occuring 0.5 Gyr or 2.0 Gyr before the SN appears too short and too long, respectively. We expect very few remaining massive stars ≥ 1 Gyr after the cessation of star formation, and therefore a negligible chance that SN1997ff could be a core-collapse SN (whose progenitors live for less than 40 Myr). An alternative history would extend the star-formation timescale, providing a small residual of ongoing star formation to boost the UV flux while allowing the rest-frame optical colors to redden to match the *IJHK* photometry. The bottom panel of Figure 7 shows such a model, with an exponential star-formation timescale of 0.3 Gyr. This model matches the observed SED at an age between 2.0 and 2.5 Gyr (with the far-UV limit favoring the older age). Normalized to the H-band magnitude of the galaxy, and assuming $\Omega_M = 0.3$, $\Omega_\Lambda = 0.7$, and $H_0 = 70$ km s^{-1} Mpc^{-1} to compute the luminosity distance, this model provides an ongoing star formation rate of 0.7 to 0.2 M$_\odot$ yr^{-1} at the time SN1997ff exploded [(2.4–10) $\times 10^{-4}$ times the initial rate]. The remaining population of massive stars should produce 0.004 to 0.001 core-collapse SNe per year (in the rest frame). We employ a more empirical route to determine the expected rate of SNe Ia due to our inability to identify conclusively their progenitor

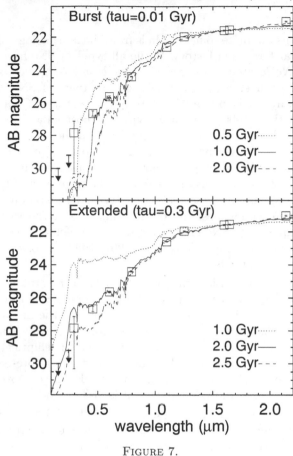

FIGURE 7.

systems. Estimates for the rate of SNe Ia at high redshift from Pain et al. (1996) yield 0.48 SNe Ia per century per 10^{10} solar blue luminosities ($H_0 = 70$ km s^{-1} Mpc^{-1}). Sullivan et al. (2000) and Kobayashi, Tsujimoto, & Nomoto (2000) predict a rise in this rate by a factor of $\sim$2 at the redshift of SN1997ff. The host galaxy luminosity is $M_B = -21.9$, and hence we expect a rate of ~ 0.07 SNe Ia per year. We thus expect the host to produce 20 to 70 times as many SNe Ia as core-collapse SNe at the time SN1997ff exploded (with the far-UV limit favoring the larger ratio), favoring its classification as a SN Ia independent of the cosmological model.

A longer timescale of star formation pushes the time of the initial burst uncomfortably close to the formation of globular clusters without significantly altering the expected production ratio of core-collapse SNe to SNe Ia.

A potentially powerful tool to discriminate between SN types comes from enlisting the observed SN data set. Both the High-z Supernova Search Team and the Supernova Cosmology Project have relied on the photometric behavior of a SN when a useful spectrum was not available (Riess et al. 1998; Perlmutter et al. 1999). The distance-independent observables of color and light-curve shape have the potential to discriminate Type Ia SNe from other SN types. As discussed in §3.1, from the observed colors and decline of SN1997ff we conclude that its photometric behavior is inconsistent with a Type II supernova at any redshift. Also, the scarcity of SNe IIn having similar photometric properties argues against SN1997ff being a SN IIn on photometric grounds. In contrast, the goodness-of-fit between SN1997ff and an empirical model of a SN Ia at $z = 1.7$ discovered a week after maximum and with a typical light-curve shape (see middle panel of Figure 4) is highly consistent with the SN Ia identification (reduced $\chi^2 = 0.5$ for ~ 10 degrees of freedom).

Based on the nature of the host galaxy (an evolved, red elliptical) and diagnostics available from the observed colors and temporal behavior of the SN, we find the most likely interpretation is that SN1997ff was of Type Ia.

4.2. Caveats

Because we have employed the SN colors to seek constraints on the SN redshift and age at discovery, both of which are strong functions of SN color, we cannot employ the colors to determine if there is any reddening by interstellar dust. The HDF-N was chosen (at high Galactic latitude) in part to minimize foreground extinction, so we assume that Milky Way reddening of SN1997ff is negligible. Similarly, given the evolved nature of the red, elliptical host, we assume negligible reddening by the host of the SN. However, the apparent consistency with past cosmological deceleration and the apparent inconsistency with contaminating astrophysical effects reported here would not be challenged by unexpected, interstellar reddening to SN1997ff. To demonstrate this conclusion, we reddened the SN by $A_B = 0.25$ mag in the rest frame and recalculated the PDF in the distance-redshift plane. As shown in Figure 6, the fit is shifted along a "reddening vector" further away from the model of astrophysical effects or cosmological non-deceleration. While we consider solutions along the reddening vector less likely, they are important to bear in mind when assessing quantitative estimates of cosmological parameters based on the previous analysis.

Our empirical model of evolution or intergalactic grey extinction (in magnitudes) is highly simplistic (i.e. linear), and consists only of the product of a constant and the redshift. Relative to the empty cosmology ($\Omega_M = 0$, $\Omega_\Lambda = 0$), this constant is chosen to be 0.3 mag per unit redshift to match the observed distances of SNe Ia at $z \approx 0.5$. The functional form of this model is the same as that derived by York et al. (2001) from the consideration of a dust-filled Universe and is shown to be valid for redshifts near unity. It also approximates the calculations of Aguirre (1999a,b). However, depending on the epoch when the hypothetical dust is expected to form, the optical depth might be expected to drop at a redshift higher than two.

Evolution is far more difficult to model and predict (Höflich et al. 1998; Umeda et al. 1999a,b; Livio 2000). For this hypothetical astrophysical effect our model is that the amount of luminosity evolution would scale with the mean age available for the growth of the progenitor system (for $z \approx 1$). While more complex parameterizations are possible, the salient feature of our simple luminosity evolution model is its monotonic increase with redshift.

Lensing by the foreground large-scale structure can alter the apparent brightness of a distant supernova (Metcalf & Silk 1999). There is a pair of galaxies in the foreground of SN 1997ff at $z = 0.56$, with a separation of $\sim 3''$ and $5.5''$ from the SN. If we assume $H_0 = 70$ km s^{-1} Mpc^{-1}, $\Omega_M = 0.3$, and $\Omega_\Lambda = 0.7$, these galaxies are found at projected distances of ~ 20 and 35 kpc in the lens plane, and have approximately L^* luminosity, with $M_V \approx -21$, implying a velocity dispersion of ~ 200 km s^{-1}. Assuming that the two galaxies have an approximately isothermal mass distribution, the resulting magnification of SN1997ff would be ~ 0.3 mag, in good agreement with the results of Lewis & Ibata (2001).

It is possible to derive a useful constraint on the *maximum likely* amplification of the SN by the closest foreground lenses by examining the shape of the host galaxy which would be stretched in the tangential direction by an amount that depends on the SN amplification. This calculation is performed in Riess et al. (2001). From this calculation we conclude that the *lack* of apparent tangential ellipticity of the host galaxy (for the degree of specific SN amplification estimated above) is not very surprising ($\sim$20% of randomly selected hosts would exhibit as little tangential stretching as seen). However, significantly greater amplifications are not very likely; 0.6 or 0.8 mag amplifications would produce galaxies with no evidence of tangential stretching only in 6% and 3% of identical ensembles, respectively.

Together, specific and stochastic lensing cases result in a possible net $\sim$0.2 mag shift in the luminosity. This value is considerably smaller than the $\sim$1 mag observed difference between the SN luminosity and that expected for astrophysical contaminants (grey dust or evolution), although it is in the direction to reduce this spread. Thus, our cosmological conclusions appear robust to lensing effects, although we cannot rule out more exotic lensing scenarios (e.g. a massive dark matter sheet amplifying the SN but not shearing the host). The challenge posed by disentangling the affects of lensing and apparent distance provide a strong impetus to collect more SNe Ia at $z > 1$ to reduce the statistical impact of such degeneracies.

5. Conclusions

1. SN1997ff is the highest-redshift SN Ia observed to date, and we estimate its redshift to be ~ 1.7.

2. The derived constraints for the redshift and distance of SN1997ff are consistent with the early decelerating phase of a currently accelerating Universe and thus are a valuable test of a Universe with dark energy. The results are inconsistent with simple evolution or grey dust, the two most favored astrophysical effects which could mimic previous evidence for an accelerating Universe from SNe Ia at $z \approx 0.5$.

3. We consider several sources of potential systematic error including gravitational lensing, supernova misclassification, sample selection bias, and luminosity calibration errors. Currently, none of these effects alone appears likely to challenge our conclusions.

I wish to express my thanks to Peter Nugent, Ron Gilliland, Marc Dickinson, Rodger Thompson, and Brian Schmidt who provided numerous contributions.

REFERENCES

AGUIRRE, A. 1999a *ApJ* **512**, 19.

AGUIRRE, A. 1999b *ApJ* **525**, 583.

ALARD, C. 2000 *A&AS* **114**, 363.

ALDERING, G., ET AL. 1998 *IAU Circ. 7046*.

BALBI, A., ET AL. 2000 *ApJ* **545**, L1.

BARBON, R., CIATTI, F., & ROSINO, L. 1979 *A&A* **72**, 287.

BARTUNOV, O. S., TSVETKOV, D. YU., & FILIMONOVA, I. V. 1994 *PASP* **106**, 1276.

BERTIN, E. & ARNOUTS, S. 1996 *A&AS* **117**, 393.

BRANCH, D., ROMANISHIN, W., & BARON, E. 1996 *ApJ* **465**, 73 (Erratum: **467**, 473).

BRUZUAL, A. G. & CHARLOT, S. 1993 *ApJ* **405**, 538.

BUDAVÁRI, T., SZALAY, A. S., CONNOLLY, A. J., CSABAI, I., & DICKINSON, M. 2000 *AJ* **120**, 1588.

CAPPELLARO, E., TURATTO, M., & FERNLEY, J. 1995, in *IUE—ULDA Access Guide No. 6: Supernovae* ESA.

CAPPELLARO, E., EVANS, R., & TURATTO, M. 1999 *A&A* **351**, 459.

CAPPELLARO, E., ET AL. 1997 *A&A* **322**, 431.

CARDELLI, J. A., CLAYTON, G. C., & MATHIS, J. S. 1989 *ApJ* **345**, 245.

CLOCCHIATTI, A., ET AL. 2000 *ApJ* **529**, 661.

COIL, A., ET AL. 2000 *ApJ* **544**, L111.

COLEMAN, G. D., WU, C.-C., & WEEDMAN, D. W. 1980 *ApJS* **43**, 393 (CWW).

COX, C., RITCHIE, C., BERGERON, E., MACKENTY, J., & NOLL, K. 1997 *NICMOS STScI Instrument Science Report* 97-07. STScI.

DE BERNARDIS, P., ET AL. 2000 *Nature* **404**, 955.

DICKINSON, M. 1999, in *After the Dark Ages: When Galaxies were Young (the Universe at 2 < z < 5)* (eds. S. Holt & E. Smith. p. 122. AIP.

DICKINSON, M., ET AL. 2001, in preparation.

FILIPPENKO, A. V. & RIESS, A. G. 2000, in *Particle Physics and Cosmology: Second Tropical Workshop*, ed. J. F. Nieves. p. 227. AIP.

FILIPPENKO, A. V., ET AL. 1992a *ApJ* **384**, L15.

FILIPPENKO, A. V., ET AL. 1992b *AJ* **104**, 1543.

GALAMA, T. J., ET AL. 1998 *Nature* **395**, 670.

GERMANY, L. M., REISS, D. J., SADLER, E. M., & SCHMIDT, B. P. 2000 *ApJ* **533**, 320.

GILLILAND, R. L., NUGENT, P. E., & PHILLIPS, M. M. 1999 *ApJ* **521**, 30 (GNP99).

GILLILAND, R. L. & PHILLIPS, M. M. 1998 *IAU Circ. 6810*.

HAMUY, M., ET AL. 1996 *AJ* **112**, 2408.

HÖFLICH, P. & KHOKHLOV, A. 1996 *ApJ* **457**, 500.

HÖFLICH, P., NOMOTO, K., UMEDA, H., & WHEELER, J. C. 2000 *ApJ* **528**, 590.

HÖFLICH, P., WHEELER, J. C., & THIELEMANN, F. K. 1998 *ApJ* **495**, 617.

IWAMOTO, K., ET AL. 1998 *Nature* **395**, 672.

JAFFE, A., ET AL. 2001; astro-ph/0007333.

KIM, A., GOOBAR, A., & PERLMUTTER, S. 1996 *PASP* **108**, 190.

KIRSHNER, R., ET AL. 1993 *ApJ* **415**, 589.

KOBAYASHI, C., TSUJIMOTO, T., & NOMOTO, K. 2000 *ApJ* **539**, 26.

LAMBAS, D. G., MADDOX, S. J., & LOVEDAY, J. 1992 *MNRAS* **258**, 404.

LEIBUNDGUT, B., ET AL. 1993 *AJ* **105**, 301.

LEWIS, G. F. & IBATA, R. A. 2001 *MNRAS* **337**, 26; astro-ph/0104254.

LIVIO, M. 2000, in *Type Ia Supernovae: Theory and Cosmology* (eds. J. C. Niemeyer & J. W. Truran). p. 33. Cambridge Univ. Press.

McLEAN, I., ET AL. 1998 *Proc. SPIE* **3354**, 566.

METCALF, R. B. & SILK, J. 1999 *ApJ* **519**, 1.

NOMOTO, K., UMEDA, H., HACHISU, I., KATO, M., KOBAYASHI, C., & TSUJIMOTO, T. 2000, in *Type Ia Supernovae: Theory and Cosmology* (eds. J. C. Niemeyer & J. W. Truran). p. 63. Cambridge Univ. Press.

OKE, J. B. & GUNN, J. E. 1983 *ApJ* **266**, 713.

PAIN, R., ET AL. 1996 *ApJ* **473**, 356.

PEEBLES, P. J. E. & RATRA, B. 1988 *ApJ* **325**, L17.

PERLMUTTER, S., ET AL. 1995 *ApJ* **440**, 41.

PERLMUTTER, S., ET AL. 1997 *ApJ* **483**, 565.

PERLMUTTER, S., ET AL. 1999 *ApJ* **517**, 565.

PETTINI, M., SHAPLEY, A. E., STEIDEL, C. C., CUBY, J.-G., DICKINSON, M., MOORWOOD, A. F. M., ADELBERGER, K. L., & GIAVALISCO, M. 2001 *ApJ* **554**, 981; astro-ph/0102456.

PHILLIPS, M. M. 1993 *ApJ* **413**, L105.

PHILLIPS, M. M., LIRA, P., SUNTZEFF, N. B., SCHOMMER, R. A., HAMUY, M., & MAZA, J. 1999 *AJ* **118**, 1766.

PHILLIPS, M. M., ET AL. 1992 *AJ* **103**, 1632.

PINTO, P. A. & EASTMAN, R. G. 2000 *ApJ* **530**, 744.

PRESS, W. H., TEUKOLSKY, S. A., VETTERLING, W. T., & FLANNERY, B. P. 1992 *Numerical Recipes in FORTRAN*, 2nd ed. Cambridge Univ. Press.

RANA, N. C. 1979 *Ap&SS* **66**, 173.

RANA, N. C. 1980 *Ap&SS* **71**, 123.

RIESS, A. G. 2000 *PASP* **112**, 1284.

RIESS, A. G., PRESS, W. H., & KIRSHNER, R. P. 1996 *ApJ* **473**, 88.

RIESS, A. G., ET AL. 1997 *AJ* **114**, 722.

RIESS, A. G., ET AL. 1998 *AJ* **116**, 1009.

RIESS, A. G., ET AL. 2000 *ApJ* **536**, 62.

RUIZ-LAPUENTE, P. & CANAL, R. 1998 *ApJ* **497**, 57.

SAHA, A., ET AL. 2001 *ApJ* **562**, 314.

SANDAGE, A. & HARDY, E. 1973 *ApJ* **83**, 743.

SAWICKI, M., LIN, H., & YEE, H. 1997 *AJ*, **113**, 1.

SCHLEGEL, E. M. 1990 *MNRAS* **244**, 269.

SCHMIDT, B. P., ET AL. 1994 *ApJ* **432**, 42.

SCHMIDT, B. P., ET AL. 1998 *ApJ* **507**, 46.

THOMPSON, R. I., STORRIE-LOMBARDI, L. J., WEYMANN, R. J., RIEKE, M. J., SCHNEIDER, G., STOBIE, E., & LYTLE, D. 1999 *AJ* **117**, 17.

TONRY, J., ET AL. 1999 *IAU Circ. 7312*.

TURATTO, M., ET AL. 2000 *ApJ* **534**, L57.

UMEDA, H., NOMOTO, K., KOBAYASHI, C., HACHISU, I., & KATO, M. 1999a *ApJ* **522**, L43.

UMEDA, H., NOMOTO, K., YAMAOKA, H., & WANAJO, S. 1999b *AJ* **513**, 861.

WILLIAMS, R. E., ET AL. 1996 *AJ* **112**, 1335.

YORK, T., ET AL. 2001, in preparation.

YUNGELSON, L. R. & LIVIO, M. 2000 *ApJ* **528** 108.

Dynamical probes of the Halo Mass Function

By C. S. KOCHANEK

Smithsonian Astrophysical Observatory, Harvard-Smithsonian Center for Astrophysics, MS-51, 60 Garden Street, Cambridge, MA 02138, USA; ckochanek@cfa.harvard.edu

We explore the relationship between the mass function of CDM halos and dynamical probes of the mass function such as the distribution of gravitational lens separations and the local velocity function. The compression of galactic halos by the cooling baryons, a standard component of modern models, leads to a feature in the distribution of lens separations near $\Delta\theta = 3\rlap{.}''0$ or in the velocity function near $v_c \simeq 400$ km s^{-1}. The two probes of the mass function, lens separations and the local velocity function, are mutually consistent. Producing the observed velocity function of galaxies or the separation distribution using standard adiabatic compression models requires more cold baryons, an equivalent cosmological density of $\Omega_{b,cool} \simeq 0.02$ compared to a total cosmological $\Omega_b \simeq 0.04$, than are observed in standard accountings of the baryonic content of galaxies, $\Omega_{b,gal} \simeq 0.006$, or our Galaxy, $\Omega_{b,Gal} \lesssim 0.015$. The requirement for a higher cold baryon density than is usually assigned to galaxies appears to be generic to models which use the standard adiabatic compression models for the transformation of the CDM halo by the cooling baryons. If real, this *dynamical baryon discrepancy* suggests either that we are neglecting half of the cold baryonic mass in standard galactic models (e.g. a MACHO, cold molecular or warm gas component), or that there is a problem with standard adiabatic compression models.

---

1 Introduction

In large part due to high resolution N-body simulations, the number density, spatial distribution, and properties of dark matter halos are well understood in models based on hierarchical clustering (e.g. Jenkins et al. 2000; Sheth & Torman 1999; Navarro et al. 1996; Moore et al. 1998). The relationship of these halos to astrophysical objects is less well understood because of the modifications to the halos produced by baryonic physics and the dependence of our search and measurement techniques on their baryonic properties. While semi-analytic models of galaxy formation (e.g. Lacey & Silk 1991, White & Frenk 1991; Cole et al. 1994, Baugh et al. 1996, Kauffman et al. 1993, 1999, Dalcanton et al. 1997; Somerville & Primack 1999; Benson et al. 2000; Cole et al. 2000†) model these effects with considerable success, the results depend on detailed, parametric models for star formation and feedback processes tuned to fit the data.

We would like to have approaches for comparing the properties of halos and astrophysical objects which minimize the dependence of the comparison on star formation and luminosity. We can generally refer to these methods as dynamical probes of dark matter halos because they focus on the observational properties of the mass distribution rather than of the luminosity distribution. This approach is well-developed for massive clusters, where there are many projects designed to determine the cosmological model and the normalization of the power-spectrum by comparing the abundance of clusters with the mass function (see the reviews by Bahcall, Donahue, Rosati & Tyson in these proceedings). Here we want to focus more on the mass scales of galaxies and on global comparisons including galaxies, groups and clusters rather than the simpler case of rich clusters. The two dynamical probes we can use to relate the halo mass function to astrophysical objects are the velocity function, the distribution of halos in their circular velocity, and the separation distribution of gravitational lenses.

† We will collectively refer to the results of semi-analytic models as SA.

Although we seek tests which avoid any dependence on the luminosity of a halo, we cannot avoid the effects of the baryons. The cooling of the baryons in the lower mass halos (i.e. galaxies) and the associated adiabatic compression of the dark matter (e.g. Blumenthal et al. 1986, SA) significantly alters the density distribution of the halo and thus the properties of any dynamical probe of the halo. While these effects are included in almost all semi-analytic models, we approach the problem from a very different viewpoint. These models are also at the center of the controversy over the consistency of the cusped dark matter profiles predicted by simulations with the observed central rotation curves of galaxies (e.g. Flores & Primack 1994, Moore 1994, de Blok et al. 2001, van den Bosch et al. 2000, 2001, Salucci 2001, also Burkert, Sancisi & Sanders in these proceedings). If there is a conflict and it is not due to an problem in the dark matter density distribution, then it must be due to a problem in either the adiabatic compression model or the assumed baryon distribution. The latter possibility is particularly interesting because we know that standard models of galaxies include only a small fraction of the available baryons.

We start in §2 with a compressed review of the models we use to determine the mass function of halos, the adiabatic compression of halos and simple cooling models for the baryons. In §3 we use these models to understand the distribution of image separations in gravitational lenses. In §4 we use the same models to study the local velocity function of galaxies and clusters. In §5 we make a final check of the consistency of the model by determining the velocity function from the gravitational lens separation distribution rather than local dynamics. Finally in §6 we outline a non-parametric approach to understanding the relation between the velocity function of galaxies and the mass function of halos. We discuss the future of dynamical probes in §7.

2. A very compressed theoretical review

We use a fixed $\Omega_0 = 0.3$ ΛCDM cosmological model with a Hubble constant of $H_0 = 67$ km s^{-1} Mpc^{-1}, a baryon density of $\Omega_b = 0.04$ and a power spectrum normalized by the abundance of rich clusters ($\sigma_8 = 0.9$). We calculated the mass function of the dark matter halos using the Press-Schechter (1974) theory combined with the Sheth & Torman (1999) fit to the results of the Virgo simulations. Where needed, we followed Kitayama & Suto (1996) and Newman & Davis (2000) in modeling the distribution of halo formation times using the extended Press-Schechter theory outlined in Lacey & Cole (1994). We neglected the problem of "halos-in-halos" (e.g. Peacock & Smith 2001, Scoccimarro et al. 2001), as it introduces considerable complexity for very modest changes in the mass function ($\sim 10\%$, White et al. 2001). In short, we assume the halo occupancy number is unity for galaxies and neglect accounting for the galaxies in clusters.

We used the Mo et al. (1998) model for the modifications to the mass distribution created by the cooling of the baryons and the adiabatic compression of the dark matter, although similar approaches are used in all SA studies. We assume that halos have the NFW (Navarro et al. 1996) density profile, $\rho \propto 1/x(1+x)^3$, where $x = r/r_s$ is the radius in units of the break radius r_s. Each halo is characterized by its virial mass M_{vir} and the concentration $c = r_{\rm vir}/r_s$. The virial mass is defined by the radius $r_{\rm vir}$ at which the enclosed density exceeds the critical density by $\Delta_c(z)$. We estimated the concentration by the mean relation $c \simeq 9(1+z)^{-1}(M_{\rm vir}/8 \times 10^{12} M_\odot)^{-0.14}$ from Bullock et al. (2000). Finally, the halo is assumed to have angular momentum J specified by its spin parameter $\lambda = J|E|^{1/2}/GM_{\rm vir}^{5/2}$ where the binding energy $|E|$ is computed using the virial theorem.

We model galaxies as exponential disks with masses of $M_d = m_d M_{\rm vir}$ and scale lengths r_d. The disk is assumed to have angular momentum $J_d = j_d J$ and this is used to determine the disk scale length (see Mo et al. 1998). Unlike Mo et al. (1998) we added a bulge modeled as a Hernquist (1990) profile with mass $M_b = m_b M_{vir}$ and an empirically estimated scale length of $0.045r_d$ from the photometry of galaxies. We usually assumed that the total specific angular momentum of the baryons was the same as that of the dark matter but that all the angular momentum is in the disk component (i.e. $j_d = m_d + m_b$ and $j_b = 0$). We use the standard Blumenthal et al. (1986) model for the adiabatic compression of the dark matter by the cooled baryons.

Finally, we used a simplified version of the Cole et al. (2000) cooling model from their semi-analytic models. We are interested in cooling because it plays an important role in determining the boundary between galaxies (where the gas can cool and form stars) and groups/clusters (where it remains hot). The Cole et al. (2000) model provides an estimate of the cooling time as a function of radius in a halo of a given virial temperature (i.e. mass), $\tau_{cool}(M, r)$, and we determine the cooled baryonic mass fraction $f_{cool}(M, z)$ at redshift z by the mass fraction inside the radius where the cooling time equals the current age, $t(z) - t_{form}(M, z) = \tau_{cool}(M, r_{cool})$. If the global baryon fraction is $(m_d + m_b)_0$ then we model the halo by an adiabatic compression model of cold baryon mass fraction $m_d + m_b = (m_d + m_b)_0 f_{cool}(M, z)$. Assuming halos start as fair samples of the universe, the global baryon fraction is $(m_d + m_b)_0 = \Omega_b / \Omega_0 = 0.13$. However, the final cold baryon fraction can be smaller, $\Omega_{b,cool} \leqslant \Omega_b$, because star formation and feedback can reheat the baryons which initially cooled (see SA).

For our scalings, the peak rotation velocity of an NFW halo of mass M is

$$v_{mod,0}(M) = 186(M/10^{12} \; M_\odot)^{0.30} \; \text{km s}^{-1} \tag{2.1}$$

which rises to

$$v_{mod}(M, m_d, \lambda) = v_{mod,0}(M) \left[1 + \frac{314 m_d^2 \Lambda^{-0.98}}{1 + 42.3 m_d \Lambda^{0.30}} \right] \tag{2.2}$$

for a halo with cold baryon fraction m_d (disk only, $m_b = 0$) and an effective spin parameter of $\Lambda = (j_d/m_d)(\lambda/\bar\lambda)$ normalized by a mean spin parameter of $\bar\lambda = 0.05$ and valid for $0 \leqslant m_d \leqslant 0.15$ and $0.02 \leqslant \lambda \leqslant 0.1$ (see Mo et al. 1998 for the effects of varying c).

3. Why don't cluster lenses exist?

One of the most striking features of surveys for gravitational lenses is that cluster lenses do not exist. This statement may seem peculiar given the enormous attention devoted to lensing by rich clusters (e.g. Tyson in these proceedings), but it is simply another facet of the fact that rich clusters loom large in our imaginations despite being exponentially rare and containing a negligible galaxy or mass fraction. The known rich cluster lenses were all found by first finding a rich cluster and then searching for lensed sources behind them.

Only surveys which examine sources to see if they are lensed probe the halo mass function, because there is a mapping between the image separation distribution $dn/d\Delta\theta$ and the halo mass function dn/dM. For example, the CLASS survey (e.g. Browne & Myers 2000, Philips et al. 2000) for lensed flat-spectrum radio sources has a nearly uniform selection function from $0''\!.3 \lesssim \Delta\theta \lesssim 15''\!.0$ and has found 18 lenses all with separations $\Delta\theta < 3''\!.0$. If we consider all 27 radio-selected lenses, there are two wider separation lenses (MG2016+112 and Q0957+561) reaching to $\Delta\theta \lesssim 6''\!.0$. Despite having the sensitivity to wide separations needed to find cluster lenses, the distribution is overwhelmingly dominated by galaxy lenses (average separations of $1''\!.5$) with a few lenses due to groups and poor clusters on larger scales (see Fig. 1).

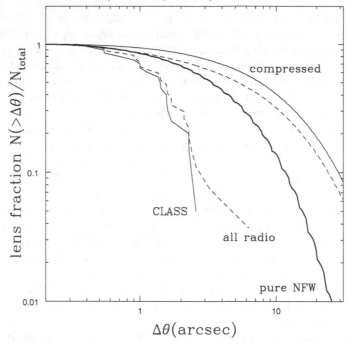

FIGURE 1. Predicted separation distributions without a cooling scale. The curves show the fraction of lenses with separations exceeding $\Delta\theta$, $N(>\Delta\theta)/N_{\text{total}}$. The observed distributions are shown by the curves labeled CLASS (for the CLASS survey lenses) and all radio (for all radio-selected lenses). The heavy solid line shows the distribution predicted by pure NFW models while the light solid (dashed) lines shows the distributions predicted by the adiabatic compression models with no bulge (a 10% baryonic mass fraction bulge).

All calculations of the separation distribution of lenses combining the halo mass function with a model density distribution for the halos catastrophically fail to explain the observed distribution of image separations (e.g. Narayan & White 1988, Kochanek 1995, Wambsganss et al. 1995, 1998, Maoz et al. 1997, Keeton 1998, Mortlock & Webster 2000, Li & Ostriker 2000, Keeton & Madau 2000, Wyithe et al. 2000). Models normalized by the abundance of rich clusters correctly find that rich cluster lenses are rare, but then grossly under predict the number of galaxy-scale lenses (see Fig. 1). A purely phenomenological approach based on the local properties of galaxies, by contrast, predicts the observed properties of the lenses well (e.g. Kochanek 1996, Keeton et al. 1998). These models have modest difficulty explaining the largest lenses found in systematic surveys ($\Delta\theta \simeq 6\rlap{.}''0$) and include no rich cluster lenses.

Keeton (1998), followed by Porciani & Madau (2000) and Kochanek & White (2001), demonstrated that the origin of the problem lay in neglecting the baryonic physics which makes the density structure of the lenses depend strongly on the mass scale. Any model based on the halo mass function which assumes that the density distributions of the halos vary smoothly and continuously with mass leads to predictions for the image separation distribution which catastrophically fail to match the data. Keeton (1998) demonstrated it for singular isothermal sphere (SIS) and NFW models, while Kochanek & White (2001) demonstrated it for the adiabatically compressed models described in §2 and illustrated in Fig. 1. The key, which can be understood self-consistently based on the adiabatic compression models described in §2 (Kochanek & White 2001), is that the lensing efficiency of a halo increases dramatically as the halo is compressed by the cooling baryons. For

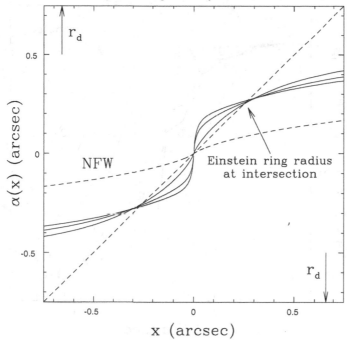

FIGURE 2. The bending angles, $\alpha(x)$, produced by a 10^{12} $M_\odot$ halo at $z_l = 0.5$ with concentration $c = 8$ for a source at $z_s = 2.0$. The dashed curve shows the bend angle for the initial sub-critical NFW halo. The solid curves show the bend angles found after the baryons cool assuming a baryonic mass fraction of $m_d + m_b = 0.05$, a spin parameter $\lambda = 0.04$, and that the disk contains all the initial baryonic angular momentum, $j_d = 0.05$. The three solid curves are for bulge-to-disk mass ratios $m_b/m_d = 0$ (shallowest rise), 0.1 and 0.2 (steepest rise) respectively. The tangential critical line of the lens (the Einstein ring) is located at the point where the (dashed) 45° line intersects the bending angle, and the radial critical line is located where a 45° line is tangent to the bending angle. An arrow points to the location of the tangential critical line of the adiabatically compressed models. The arrows on either side of the figure indicate the (angular) disk scale length r_d at the model redshift.

example, Fig. 2 shows the change in the deflection profile (bend angle) for a 10^{12} $M_\odot$ halo at $z_l = 0.5$ with a source at $z_s = 2$ between the initial NFW model and the final adiabatically compressed model with mass fraction $m_d + m_b = 0.05$ in cold baryons. Where the initial NFW halo is sub-critical and unable to generate multiple images, the compressed halo is super-critical. The final cross section depends strongly on the details of the baryon distribution, as the models with bulge-to-disk mass ratio of $m_b/m_d = 20\%$ are significantly better lenses than those without a bulge component.

Keeton (1998) and Porciani & Madau (2000) phenomenologically solved the problem of determining the global separation distribution by breaking the mass function into a high mass (cluster) distribution modeled by standard NFW profiles and a low mass (galaxy) distribution modeled using the local properties of galaxies rather than the theoretical mass function. In order to fit the observed separation distribution, the break had to be located at a cooling mass scale near $M_c = 10^{13}$ $M_\odot$. Given the difficulty in performing *ab initio* calculations for the final density structure of galaxies (as illustrated by the strong dependence of the bend angles on the bulge mass fraction in Fig. 2), such phenomenological methods are likely to be more quantitatively useful than current attempts at ab initio calculations.

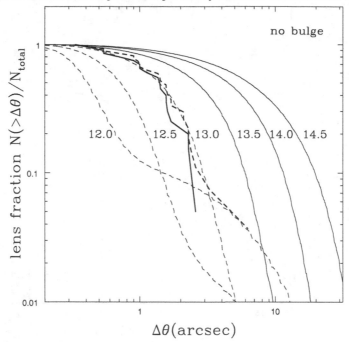

FIGURE 3. Predicted separation distributions with a cooling scale for models without a bulge and cold baryon fraction $m_d + m_b = 0.05$ in halos below a cooling mass scale M_c. The curves are labeled by $\log M_c/M_\odot$. The heavy solid (dashed) curve shows the observed distribution of the CLASS (radio-selected) lenses.

While recognizing this limitation on the quantitative use of current theoretical models for the final density structure of galaxies, Kochanek & White (2001) used the theoretical models to illustrate and isolate the physics needed to produce the observed separation distribution from the halo mass function. They first assumed that for halos less massive than a cooling mass scale M_c, baryonic mass fraction $m_d + m_b = 0.05$ of the halos cools. Fig. 3 shows the predicted separation distributions as a function M_c. Recall, from Fig. 1, that if either all (high M_c) or no (low M_c) halos cool, then we cannot match the observations. Only for a cooling scale $M_c \sim 10^{13}$ $M_\odot$ can we produce a sharp break in the separation distribution on the $\Delta\theta = 3''0$ scale of the observations. The exact scale depends on the density structure of the cooled halos, with M_c decreasing from 10^{13} $M_\odot$ to 5×10^{12} $M_\odot$ when we add a $m_b/m_d = 0.1$ bulge. Interestingly, cosmological hydrodynamic simulations also find that the cooled baryon fraction reaches 50% on mass scales near 10^{13} $M_\odot$ (e.g. Pearce et al. 1999). The required cooling mass scale also depends on the cold baryon fraction, with M_c decreasing as $\log_{10} M_c/ M_\odot \simeq 13.6 - (m_d + m_b)/0.15$ when the cold baryon fraction $m_d + m_b$ increases. The Kolmogorov-Smirnov (K–S) test probability of fitting the observed separation distribution is shown in Fig. 4 for a range of cold baryon fractions.

These two parameters, the cooling mass scale and the cold baryon fraction, are not independent. The cooling mass scale M_c at any epoch is the mass scale where the cooling time is roughly equal to the average age of the halos. Fig. 4 also shows that for the current epoch, the two time scales are equal near $M_c = 10^{13}$ $M_\odot$ for $m_d + m_b = 0.05$. A lower cold baryon fraction requires a mass scale M_c for which there is insufficient time for the baryons to cool. A higher cold baryon fraction requires mass scales where there is too much time for the baryons to cool. Since (to first order!) cooling physics

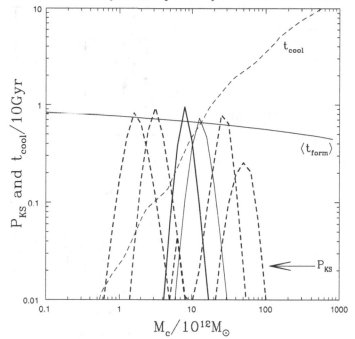

FIGURE 4. The Kolmogorov-Smirnov probability, P_{KS}, of fitting the observed separation distribution of CLASS lenses as a function of the cooling mass scale M_c. The heavy (light) solid curves indicated by the arrow show the K-S probability for models with $m_b + m_d = 0.05$ without (with) a $m_b/m_d = 0.10$ bulge. The heavy dashed curves show the K-S probabilities for models with lower ($m_b + m_d = 0.01$ and 0.02) or higher ($m_b + m_d = 0.10$ and 0.20) baryon fractions where the optimal cooling mass decreases as the baryon fraction rises. The light dashed curves show the cooling time in units of 10 Gyr for the radii enclosing 50% of the baryonic mass for the standard model. The light solid line shows the time since the average formation epoch ($\langle t_{\text{form}} \rangle$) in units of 10 Gyr assuming $h = 0.67$.

combined with the baryonic mass fraction determines M_c, the fundamental physical parameter leading to the observed separation distribution is the cosmological baryon density Ω_b. If all baryons cooled and remained cold, then there would be no ambiguities to this statement. However, star formation and feedback can reheat large fractions of the baryons even if they cool initially (see SA), so that the cosmological density in cold baryons, $\Omega_{b,\text{cool}}$, can be significantly less than the total density in baryons Ω_b. In our simple models we assume that halos start with (eventually cold) baryon mass fraction $(m_d + m_b)_0 = \Omega_{b,\text{cool}}/\Omega_m \lesssim \Omega_b/\Omega_m = 0.13$ and that the fraction which has cooled and compressed a halo of mass M at redshift z is $m_d + m_b = (m_d + m_b)_0 f_{\text{cool}}(M, z)$ where the simple model of the cooling function outlined in §2 determines $f_{\text{cool}}(M, z)$. Fortunately, the estimates of $\Omega_{b,\text{cool}}$ depend very weakly on changes in f_{cool}, the lens sample used, or the addition of a small bulge component.

The K–S test probability of fitting the observed separation distribution as a function of the cold baryon density $\Omega_{b,\text{cool}}$ is shown in Fig. 5. With little sensitivity to the details, models with $0.015 \lesssim \Omega_{b,\text{cool}} \lesssim 0.025$ agree with the data. While the preferred range is less than the total baryon density $\Omega_b = 0.04$ in the input cosmology, it significantly exceeds the estimates of $0.0045 \lesssim \Omega_{b,\text{cool}} \lesssim 0.0068$ for the cold baryon fraction (stars, cold gas and stellar remnants) in local galaxies by Fukugita, Hogan & Peebles (1998). This is a weighted average over all galaxies rather than simply that of massive galaxies, but the conflict is probably present even for massive galaxies like the Milky Way. Models

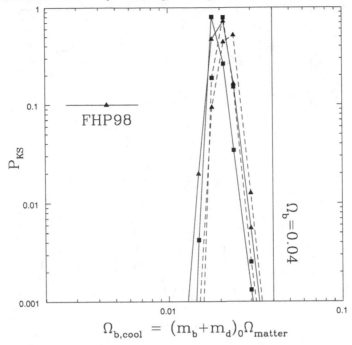

$$\Omega_{b,cool} = (m_b + m_d)_0 \Omega_{matter}$$

FIGURE 5. Kolmogorov-Smirnov test probability of fitting the separation distribution of CLASS lenses as a function of $\Omega_{b,cool}$. The squares (triangles) indicate models with no bulge (with a $m_b/m_d = 0.1$ bulge), and the solid (dashed) lines correspond to fitting the CLASS lenses (all radio lenses). The point with horizontal error bar is the estimate by Fukugita, Hogan & Peebles (1998) for the cold baryon (stars, remnants, cold gas) content of galaxies. The vertical line marks the total baryon content in the concordance model.

of the Galaxy by Dehnen & Binney (1998) find a baryonic mass fraction $m_d + m_b \simeq$ 0.08 corresponding to $\Omega_{b,cool} \simeq 0.024$ at 100 kpc. As we increase the radius we include additional dark matter but no new baryons, so the baryon fraction drops to $0.008 \lesssim$ $\Omega_{b,cool} \lesssim 0.012$ for a halo with an outer extent of 200–300 kpc. While the difference is smaller, it suggests the existence of a *dynamical baryon discrepancy*, in which the cold baryon fraction required to explain the observed dynamical properties of galaxies exceeds standard accountings for the cold baryons in galaxies.

4. The Local Velocity Function

The distribution of gravitational lens image separations is the cleanest dynamical probe of the halo mass function because the image separation is directly related to the underlying halo density distribution and because lensing selects halos without any dependence on the luminosity of the baryons. Unfortunately, while the number of known lenses is growing relatively rapidly, the overall size of the sample is limited. Our alternative dynamical probe of the mass function is the local velocity function of galaxies and clusters. The local velocity function has the opposite problems from the lenses, with far lower statistical uncertainties but far higher systematic uncertainties. First, selection methods for galaxies and clusters depend on the luminosity and surface brightness of the halo. Second, the selection methods for galaxies and groups/clusters are inhomogeneous, leading to significant systematic difficulties when assembling a global velocity function incorporating both. Third, local dynamical probes of galaxies and clusters have many more ambiguities than gravitational lens image separations because they are only indirectly related to the mass distribution. This problem is worst for galaxies where there are significant systematic uncertainties in the relationship between stellar kinematics and the underlying mass distribution.

Most estimates of the local velocity function of galaxies have been made because it is an essential element in estimates of gravitational lens cross sections, separation distributions, and related statistics (e.g. Turner et al. 1984, Fukugita & Turner 1991, Kochanek 1993, 1996; Maoz & Rix 1993; Falco et al. 1998; Helbig et al. 1999). Cole & Kaiser (1989) pointed out that the velocity function can also be used to test models of galaxy formation, the power spectrum and cosmology directly. This motivated estimates of the local velocity function by Shimasaku (1993) and Gonzalez et al. (2000) and further theoretical investigations by Sigad et al. (2000), Newman & Davis (2000), and Bullock et al. (2001). All derivations of the galaxy velocity function use the Faber-Jackson (1976) and Tully-Fisher (1977) relations between luminosity and velocity to perform a variable transformation from a locally measured luminosity function into the velocity function. Far more effort has focused on determining the mass or velocity (temperature) function of groups and clusters (see the reviews by Rosati & Donahue in these proceedings) and given the greater familiarity of these results we will not review them in any detail. It is difficult to merge the two, to produce a global view of the mass function, because of the problems in obtaining complete, well-understood samples of groups and then determining their masses or velocities (e.g. Mahdavi et al. 2000).

Existing observational estimates of the velocity function of galaxies suffer from (at least!) five systematic problems. First, deriving a velocity function requires the luminosity function of galaxies by type since the kinematics of pressure-supported early-type and rotation-supported late-type galaxies are very different. Unfortunately, there are large systematic differences between morphologically-typed and spectrally-typed luminosity functions. However, in Kochanek et al. (2001b) we find clear evidence that current spectrally-typed luminosity functions have internal inconsistencies which make the morphologically-typed surveys the better option at present. Second, the luminosity functions and the kinematic relations used to construct the velocity function are derived on different magnitude scales, leading to systematic shifts in the velocity scale. Third, by separating the derivations of the luminosity function and the kinematic relations, covariances which are important to the uncertainties in the velocity function are lost. Fourth, the classification method used for the luminosity function (morphological or spectral) is not the same as the kinematic classification (pressure or rotation supported) used to define the kinematic relations. The galaxies found in kinematic samples do not even represent fair samplings of the classifications used to construct the luminosity function as they are dominated by E and Sb/Sc galaxies with few S0/Sa galaxies. Fifth, the kinematic velocities have specific observational definitions that are not easily translated into theoretically calculated quantities. For example, the central stellar velocity dispersion of an early-type galaxy is non-trivially related to the peak circular velocity of a model halo.

In Pahre et al. (2001) we took three steps to reducing these systematic uncertainties. First, we started from a large, local infrared luminosity function (Kochanek et al. 2001a) derived from the 2MASS (Skrutskie et al. 1997) survey. The galaxies were morphologically classified and the classifications are internally consistent and consistent with other local, morphologically classified samples (Kochanek et al. 2001b). We then matched galaxies in the 2MASS sample to several modern kinematic surveys to construct Faber-Jackson and Tully-Fisher relations in the same magnitude system as was used to derive the luminosity function. We then combined the luminosity function and the kinematic relations including the full variable covariances of the functions to determine the velocity function.

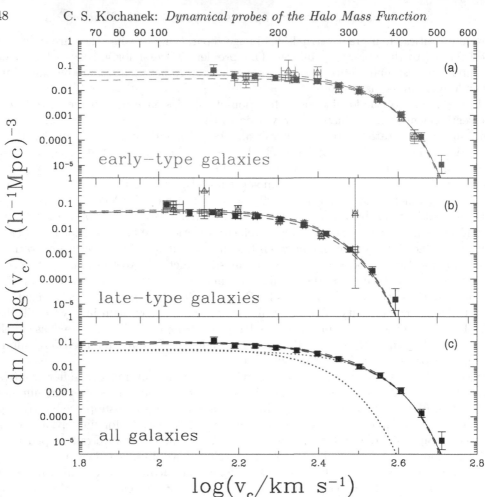

FIGURE 6. (a) The early-type galaxy velocity function. The curves show our parametric estimates of the velocity function. The solid curve is our standard estimate with the galaxy type boundary at $T = -0.5$, while the upper (lower) curves show that the number of low velocity early-type galaxies is strongly affected by shifts in the type boundary to $T = +0.5$ (-1.5). The points show the three different non-parametric estimates of the velocity function derived from the binned, non-parametric estimate of the luminosity function. The solid squares use the analytic Faber-Jackson relation to transform both the magnitude and the density, the open squares use it only to transform the density, and the open triangles do not use it at all. (b) The late-type galaxy velocity function. The curves and symbols are the same as in (a). (c) The total galaxy velocity function. The curves are the same as in (a) and (b), but now calculated for all galaxies simultaneously. The individual early- and late-type galaxy contributions to the velocity function are plotted as dotted lines.

We also explored non-parametric models for the velocity function where we minimized the use of functional forms for the luminosity function or the kinematic relations in favor of the raw, binned data. Fig. 6 shows the resulting velocity functions, and Fig. 7 compares our velocity function to other derivations (from Gonzalez et al. 2000) and to the standard result from gravitational lens statistics (Falco et al. 1998). While our result has reduced systematic errors compared to earlier derivations, we have not eliminated problems created by the difference between morphological and kinematic types or the problems in relating observed and theoretical velocities.

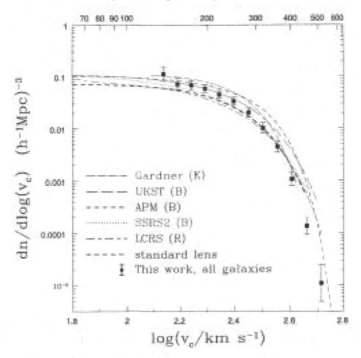

FIGURE 7. The total galaxy velocity function compared to previous estimates from Gonzalez et al. (2000) and Falco et al. (1998). The standard lens model can be offset in velocity because of differences in the observational method (see §5).

Our theoretical model, tuned to reproduce the distribution of gravitational lens separations in §3, also estimates the peak circular velocity as a function of mass, as shown in Fig. 8. Below the cooling mass scale, the baryons compress the halos and shift the mass-velocity relation $v_c(M)$ upwards. The amount of the shift depends on both the assumed baryon distribution (as illustrated by comparing models with bulge-to-disk mass ratios of $m_b/m_d = 0$ and 0.1) and redshift. With added physics (such as the reheating mechanisms in the SA models), the simple power-law structure below the cooling mass scale changes. We can use the mass-velocity relation to derive the velocity function from the mass function as a simple variable transformation (eqn. 6.2 with $p_g(M) = 1$) with the results shown in Fig. 9. We used three $v_c(M)$ relations: the virial velocity of the NFW halos, the peak circular velocity of the NFW halos, and the $v_c(M)$ relation produced by the model which best fit the gravitational lenses. We also extended the observed velocity function by adding a crude estimate for the contribution of groups and clusters to the local velocity function. We estimated the cluster contribution from the Blanchard et al. (2000) X-ray temperature function combined with the Wu et al. (1999) relation between the temperature and the velocity dispersion (a simple thermodynamic conversion is nearly identical) truncated to $T > 0.5$ keV. This produces the small kink in the estimate near $v_c = 400$ km s^{-1}. On cluster scales the models and the data agree relatively well, although a modestly lower normalization for the mass function would fit the data better. On galaxy scales, the models without the compression by the baryons grossly under predict the density of halos with observed velocities 200 km s$^{-1} \lesssim v_c \lesssim 400$ km s^{-1} while the models with adiabatic compression match the density relatively well.

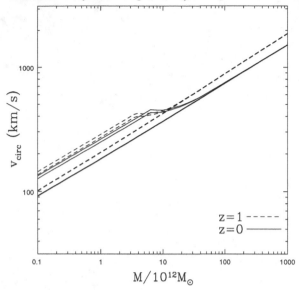

FIGURE 8. The global relation between mass and circular velocity at redshifts zero (solid) and unity (dashed). The heavy curves show the peak circular velocity of the NFW model. The light curves show the peak circular velocity including the baryonic cooling and adiabatic compression from the $\Omega_{b,\text{cool}} = 0.018$ model. The upper light curve is the model with no bulge component ($m_b/m_d = 0$) and the lower light curve is the model with a bulge ($m_b/m_d = 0.1$). The bulge slightly reduces the peak rotation velocity (because it increases the angular momentum per unit mass of the disk) while making the rotation curve flatter.

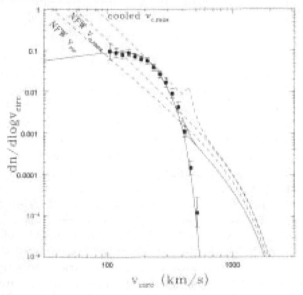

FIGURE 9. The velocity function $dn/d\log v_c = (dn/d\log M)|d\log M/d\log v_c|$. The solid curves show the local velocity function of galaxies (low v_{circ}) and clusters (high v_c) and their sum. The points are the non-parametric velocity function of galaxies. From bottom to top, the dashed curves show the velocity functions derived using dn/dM and the NFW virial velocity (labeled NFW v_{vir}), the peak circular velocity of the NFW rotation curve (labeled NFW $v_{c,\text{max}}$) and the peak circular velocity of the adiabatically compressed model (labeled cooled $v_{c,\text{max}}$). We used the $\Omega_{b,\text{cool}} = 0.018$ model with no bulge.

Our model clearly has problems at low velocity and at the juncture between galaxies and groups/clusters. These are not apparent in the models for the gravitational lenses because the lens cross sections combined with the angular selection functions strongly suppress the contributions from halos with $v_c \lesssim 150$ km s^{-1}, and because the separation distribution smoothes the velocity function by an average over lens redshift. The peak in the model distribution near 400 km s^{-1} is an artifact of the flat slope of the $v_c(M)$ curves (see Fig. 8). The $v_c(M)$ relation has to be multi-valued in this region with a diminishing probability of forming a compressed galaxy halo and a rising probability of forming an uncompressed group halo (we will explore this further in §6). Gonzalez et al. (2000), who made a similar comparison to the velocity function of galaxies based on the Somerville & Primack (1999) semi-analytic models, had very similar problems. Even though they used a model where feedback from star formation varied the cold baryon fraction to maximize the compression at $v_c \sim 250$ km s^{-1} and to reduce it for higher and lower velocity halos, their model velocity function more closely resembles our theoretical curves than the observations.

5. Deriving the Velocity Function From the Lens Separations

As a final check of the consistency of the model, the distribution of lens separations and the velocity function, we can estimate the velocity function directly from the distribution of lens separations so that the comparison does not depend on our theoretical model from §2. We will compare only the shapes of the velocity function and not the absolute normalization (number per comoving Mpc) since the normalization introduces the uncertainties in the absolute numbers of gravitational lenses found by a survey. We assume that lenses can be modeled as singular isothermal spheres (SIS), which is broadly consistent with both lensing and dynamical data on the relevant scales (see Cohn et al. 2001), so that the observed image separation is a simple function of the circular velocity v_c and the lens-source/observer source distance ratio, $\Delta\theta = \Delta\theta_0 (v_c/v_0)^2 D_{LS}/D_{OS}$ where $\Delta\theta_0 = 4\pi(v_0/c)^2$ sets an arbitrary velocity scale for the calculation. In any flat cosmology, the normalized image separation distribution is

$$\frac{1}{\tau}\frac{d\tau}{d\Delta\theta} = \frac{\Delta\theta^2 S(\Delta\theta)}{\Delta\theta_0} \left[\int_0^1 x^2 dx \int_0^\infty dv \Delta\theta^2 S(\Delta\theta)\frac{dn}{dv} \right]^{-1} \int_{v_{min}}^\infty dv \frac{v_0^2}{v^2}\frac{dn}{dv}\left(1 - \frac{v_0^2}{v^2}\frac{\Delta\theta}{\Delta\theta_0}\right)^2$$

$$(5.1)$$

where $v_{min} = v_0(\Delta\theta/\Delta\theta_0)^{1/2}$, $0 \leqslant S(\Delta\theta) \leqslant 1$ is the survey selection function for finding a lens of separation $\Delta\theta$, $x = D_{OL}/D_{LS}$, and dn/dv is the velocity function. We can non-parametrically determine the velocity function using a variant of the step-wise maximum likelihood (SWML) luminosity function estimation method of Efstathiou et al. (1988). We approximate dn/dv by a series of bins with density n_i and then maximize the likelihood of finding the observed lens separations while holding the number of lenses in the bin centered on $v_c = 300$ km/s fixed to unity. The latter constraint corresponds to ignoring the absolute comoving density of the lenses and considering only the shape of the velocity function. Eqn. (5.1) holds for any flat cosmological model.

The results for three lens samples are shown in Fig. 10. Sample A consists of the 20 lenses found in surveys based 8 GHz VLA A-array maps of flat-spectrum radio sources by the CLASS and PMN (Winn et al. 2000) surveys. Sample B, with 27 lenses, adds the remaining radio-selected lenses. Sample C, with 46 lenses, adds the optically-selected quasar lenses. The model for the angular selection function is adjusted for the angular sensitivity of the survey which found each lens. As we move from Sample A to C we trade increasing systematic uncertainties for decreasing statistical uncertainties, although

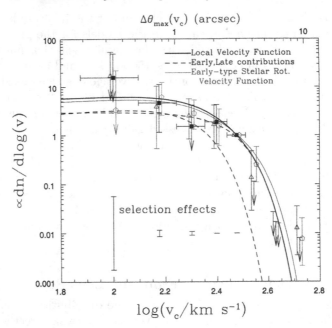

FIGURE 10. The velocity function estimated from gravitational lenses. The solid points are derived using only the flat-spectrum radio surveys, the triangles include the MIT-Greenbank survey, and the pentagons include quasar lenses. The triangles and pentagons are slightly offset in velocity to make them more visible. The horizontal error bars show the width of the velocity bins. The distributions are normalized by the logarithmic density at $v_c = 300$ km s^{-1}. The solid (dashed) line shows our locally estimated total (early-type and late-type) velocity function. The error ranges labeled "selection effects" show the effects of plausible uncertainties in the angular selection function $S(\Delta\theta)$. The scale at the top of the figure shows the maximum image separation produced by a lens with circular velocity v_c. The mean separation is one-half the maximum.

we find that the derived velocity functions are mutually consistent given the statistical uncertainties. Only the lowest velocity bin, centered at $v_c = 100$ km/s and corresponding to an average image separation of only $\langle\Delta\theta\rangle = 0\rlap{.}''15$, has a significant sensitivity to plausible errors in the models for the angular selection function. Be warned that the error bars in Fig. 10 are even more highly correlated than similar figures derived for luminosity functions of galaxies!

If we normalize our estimate of the velocity function of galaxies at the same velocity scale and superpose it on that of the lenses, the two distributions are remarkably similar (see Fig. 10) They have the same flat low-velocity slope as a function of $\log(v_c)$ and an exponential cutoff on the same velocity scale. Only in the highest velocity bin, whose density is driven by the need to produce the widest separation Q 0957+561 lens with $\Delta\theta = 6\rlap{.}''1$, does the velocity function of the lenses show a clear deviation from that of galaxies even though the lenses represent the global velocity function rather than that of the galaxies alone. The small amplitude of the deviation is another illustration of the enormous impact of the baryons on the dynamical structure of halos which we discussed in §3. At low velocities the two distributions have the same flat slope, rather than the steeply rising slope of our estimates from the mass function. Note, however, that neither observational sample has significant data on velocity scales $v_c \lesssim 100$ km s^{-1} where the deviations from the predictions begin to diverge rapidly.

6. A non-parametric description of the Velocity Function

Our theoretical model is a gross oversimplification. It agrees with our two observational probes on the mass scales corresponding to massive galaxies and clusters, but has problems on the mass scales of groups and fails badly for low mass galaxies. These problems can be partially rectified by more sophisticated models (e.g. SA) which allow for more complicated variations in the cold baryon fraction with halo mass. In this section we outline a general, non-parametric approach to understanding the relationship between the mass function of halos and the velocity function of galaxies which we can use to characterize the problem without the complications of a full semi-analytic model.

Our starting point is that in all models, no matter their complexity, the halo mass function and the galaxy velocity function are related by the probability $P(v_c|M)$ that a halo of mass M forms a detectable galaxy with circular velocity v_c,

$$\frac{dn}{dv_c} = \int_0^\infty \frac{dn}{dM} P(v_c|M) dM. \tag{6.1}$$

The conditional probability, which need not integrate to unity, includes the effects of all parameters governing the formation of galaxies such as the spin parameter λ, the collapse redshift, the halo merger history and its environment. If $P(v_c|M)$ is dominated by a sufficiently narrow ridge, so that it is reasonable to associate a characteristic velocity with halo mass, then we can approximate the integral (6.1) by

$$\frac{dn}{dv_c}(v_c(M)) = p_g(M) \left| \frac{dv_c(M)}{dM} \right|^{-1} \frac{dn}{dM} \tag{6.2}$$

where the unknown two-dimensional function is replaced by two one-dimensional functions with simple physical meanings. The first, $p_g(M)$, is the probability that a halo of mass M forms a galaxy included in the velocity function, and the second, $v_c(M)$, is the circular velocity of the resulting galaxy. As a mathematical derivation we must assume that the fractional spread in velocity at fixed mass, $\sigma_v(M)/v_c(M)$, is small. Since the equivalent fraction at fixed luminosity is indeed small, and changing from mass to luminosity presumably raises rather than lowers the dispersion in galaxy properties, it seems likely that the expansion is justified. This model differs from that used in §3 and §4 where $p_g(M) \equiv 1$. Its main drawback is that it neglects "halos-in-halos" or the halo multiplicity function (e.g. Peacock & Smith 2000, Soccimarro et al. 2001), although this should be a modest perturbation in the accounting for galactic halos ($\sim 10\%$ in the numerical simulations of White et al. (2001). These two functions are implicitly included in semi-analytic models. For example, Gonzalez et al. (2000) fit the cold baryon fraction in the Somerville & Primack (1999) models by $m_d(x) = 0.1(x - 0.25)/(1 + x^2)$ for $x = v_{mod,0}(M)/200$ km s$^{-1} > 0.25$ (see eqn. 2.1) to estimate $v_c(M)$ for the model. The velocity function predicted by this model cannot, however, reproduce the local velocity function of galaxies, as discussed in §4 (see Fig. 9).

Our decomposition of the problem into the formation probability $p_g(M)$ and the mass-velocity relation $v_c(M)$ allows us to explore the problem in a model independent fashion. Unfortunately, the solution is not unique because we must determine both $p_g(M)$ and $v_c(M)$ from only one function. Fortunately, the two unknown functions should obey several constraints. First, if the normalization of the mass function is correct, $0 \leqslant p_g(M) \leqslant 1$. We can have $p_g(M) > 1$ only if we interpret it as a normalization error in the mass function or if the halo multiplicity function on galactic mass scales differs significantly from unity. Second, cooling baryons should only increase the halo circular velocity, $v_c(M) \geqslant v_{mod,0}(M)$. Third, the compression must be bounded by models

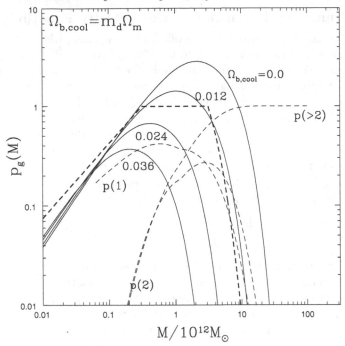

FIGURE 11. The galaxy formation probability, $p_g(M)$, required for models with constant cold baryon fractions. The curves are labeled by their cold baryon density, $\Omega_{b,\text{cool}} = m_d \Omega_m$. The upper curve is for halos with no cooled baryons ($\Omega_{b,\text{cool}} = 0$) and the lowest curve is for halos with $\Omega_{b,\text{cool}} \simeq \Omega_b$. The heavy dashed curve is the standard model from our attempt to non-parametrically adjust $v_c(M)$ in order to maximize the mass range over which $p_g(M) = 1$. The light dashed curves show the probability of forming one ($p(1)$), two ($p(2)$) or at least two ($p(> 2)$) galaxies as a function of halo mass in one of the halo multiplicity models of Scoccimarro et al. (2001). The Scoccimarro et al. (2001) models are low because they underestimate the comoving density of galaxies (see text).

in which all the available baryons have cooled, $v_c(M) \leqslant v_{\text{mod}}(M, m_d = \Omega_b/\Omega_m, \lambda)$. Note that this upper bound is more model dependent than the lower bound.

Suppose that the standard adiabatic compression models are correct and that the dimensionless parameters (cold baryon fraction, spin parameter $\cdots$) are the same for all galaxies. These parameters fix $v_c(M)$, allowing us to calculate the galaxy formation probability required to produce the galaxy velocity function from the halo mass function. Fig. 11 shows the implied $p_g(M)$ for various cold baryon fractions ($m_d = 0$, 0.04, 0.08 and 0.12) and a fixed spin parameter $\lambda = \bar{\lambda}$). At low mass, $p_g(M) \sim M$ is needed to match the steep slope of the mass function to the shallow slope of the velocity function, and at high mass it has the exponential cutoff of the velocity function. The peak probability and the corresponding mass scale decrease systematically as we raise the cold baryon fraction. If the compression is too small ($m_d \lesssim 0.04$ or $\Omega_{b,\text{cool}} \lesssim 0.01$), we must increase the normalization of the halo mass function (i.e. the power spectrum) in order to avoid a region with $p_g(M) > 1$. When the compression of the halos is large ($m_d \gtrsim 0.08$ or $\Omega_{b,\text{cool}} \gtrsim 0.02$), either the velocity function is incomplete or we must lower the normalization of the mass function.

While these models clearly simplify the structure of the mass-velocity relation $v_c(M)$ by using models with fixed dimensionless parameters, they do not greatly exaggerate the dominant role of the formation probability in producing the observed shape of the

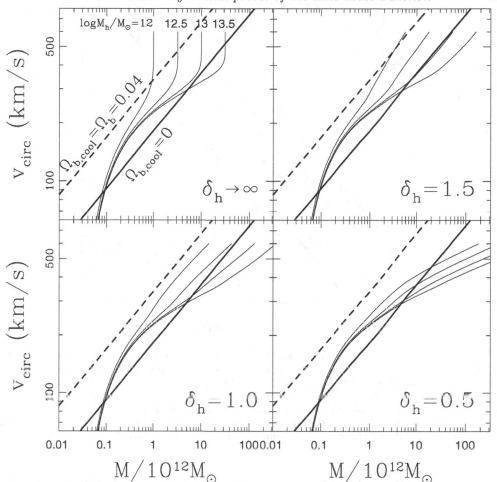

FIGURE 12. The effect of the high mass cutoff in the galaxy formation probability on the mass-velocity relation. The four light curves in each panel show the mass-velocity relations $v_c(M)$ needed to produce the observed velocity function given high mass cutoffs in the formation probability of $\log M_h/M_\odot = 12.0$, 12.5, 13.0 and 13.5 (from left to right). The different panels show the effect of changing the cutoff exponent, with values of $\delta_h \to \infty$ (top left), 1.5 (top right), 1.0 (lower left) and 0.5 (lower right) covering the range from an infinitely sharp cutoff to a fairly shallow cutoff. The heavy curves show the permitted range for $v_c(M)$ with a lower boundary set by the circular velocity of uncompressed halos (the heavy solid line) and the upper boundary set by the circular velocity of halos in which all the available baryons have cooled (the heavy dashed line). The upper bound, which assumes $\lambda = \bar{\lambda} = 0.05$ and $j_d = m_d$, is significantly more model dependent than the lower bound.

velocity function. The requirement that a physical mass-velocity relation be bounded by the zero compression and maximal compression models prevents us from shaping $v_c(M)$ to produce the velocity function with the formation probability held constant over much more than one decade in mass. We can illustrate this by using a parametric model for $p_g(M)$ with a constant formation probability $p_g = p_0$ between $M_l < M < M_h$, a power-law $p_g = p_0(M/M_l)^{\delta_l}$ below M_l, and an exponential $p_g = p_0 \exp(1 - (M/M_h)^{\delta_h})$ above M_h. For any set of parameters for $p_g(M)$ we can derive the $v_c(M)$ required to produce the velocity function.

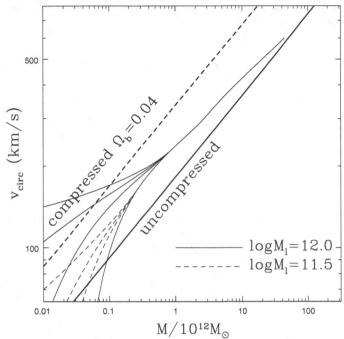

FIGURE 13. The effect of the low mass cutoff in the galaxy formation probability on $v_c(M)$. The light solid and dashed lines show the $v_c(M)$ relations needed to produce the observed velocity function given low-mass cutoffs at $\log M_l/M_\odot = 11.5$ (dashed) and 12.0 (solid) for power law exponents of $\delta_l = 1$ (top), 3/4, 1/2, and 0 (bottom). The case $\delta_l = 0$ corresponds to having no low-mass cutoff and the two M_l cases overlap. The upper cutoff is fixed to $M_h = 10^{12.5}\ M_\odot$ with $\delta_h = 1.0$. The heavy solid and dashed lines are the lower and upper bounds corresponding to the uncompressed and the maximally compressed limits of the model.

Figs. 12–14 illustrate the effect of adjusting the structure of the formation probability on the mass-velocity relation $v_c(M)$. We start with the high-mass cutoff (M_h and δ_h) since we are certain it exists in order to create the distinction between galaxies and groups/clusters. Physical models, where $v_c(M)$ stays within its physical bounds, require a slightly blurred boundary with $\delta_h \simeq 1$ and $M_h \simeq 10^{12.5}\ M_\odot$ (which we adopt as our standard). Steeper slopes δ_h allow modestly higher mass scales M_h but the mass where $p_g(M) = 1/2$ stays roughly in the range $10^{12.5}\ M_\odot$ to $10^{13}\ M_\odot$. The group velocity function is set by $1 - p_g(M)$ with a velocity set by that of uncompressed halos, $v_c(M) = v_{\mathrm{mod},0}(M)$. It is the lack of this multi-valued region in our models from §2 which leads to the "kink" in our model of the velocity function near $v_c \sim 400$ km s^{-1} (see Fig. 9).

Without a low-mass cutoff in the formation probability, $v_c(M)$ would have to rise exponentially with mass in order to convert the steep slope of the mass function into the shallower slope of the velocity function (see Fig. 9). Since this rapidly requires velocities well below that of the halos, we are forced to introduce a low-mass cutoff (M_l and δ_l) to the formation probability. The effect of changing various parameters is shown in Fig. 13, and we adopt $M_l = 10^{11.5}\ M_\odot$ and $\delta_l = 3/4$ as our standard model. All models which allow $v_c(M) > v_{\mathrm{mod},0}(M)$ down to 30 km s^{-1} have a formation probability near 50% at $10^{11}\ M_\odot$ and 10% near $10^{10}\ M_\odot$. We introduced these models for $p_g(M)$ to see if we could produce a broad mass range in which the formation probability was unity. Fig. 11 superposes the standard model we derived on the models for $p_g(M)$ found without any complicated structure to $v_c(M)$. The structure of the functions is nearly identical—the

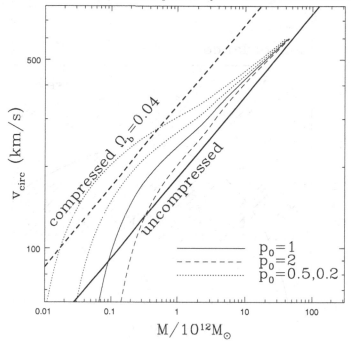

FIGURE 14. The effect incompleteness or normalization errors on $v_c(M)$. The light curves $v_c(M)$ for $p_0 = 2$ (dashed), $p_0 = 1$ (solid), $p_0 = 1/2$ (dotted) and 1/5 (dotted) using the standard high mass cutoff. The heavy solid and dashed lines are the lower and upper bounds corresponding to the uncompressed and the maximally compressed limits of the model.

best we can manage is to keep the formation probability unity over one decade in halo mass. From this we conclude that the structure of the velocity function is dominated by the probability $p_g(M)$ of forming (or finding) a galaxy rather than variations in the relationship between halo mass and galaxy circular velocity $v_c(M)$.

Finally, in Fig. 14 we explore the effects of varying p_0, which models either errors in the normalization of the mass function or the completeness of the survey underlying the velocity function. We use the standard high mass cutoff and no low mass cutoff. For $p_0 < 1$, which corresponds to making the velocity function survey incomplete or lowering the normalization of the mass function, the velocity corresponding to a given halo rises, and for $p_0 > 1$, which corresponds to raising the normalization of the mass function, the velocity falls. From this we can infer that the velocity function galaxy sample cannot be massively incomplete and the halo multiplicity function cannot differ greatly from unity.

All these conclusions are very similar to the results from attempts to estimate the halo occupation numbers needed to match the halo distributions found in simulations to the observed distribution of galaxies (e.g. Peacock & Smith 2000, Scoccimarro et al. 2001). For example, Fig. 11 also shows the probability of a halo forming one, two or many galaxies in one of the models by Scoccimarro et al. (2001) based on semianalytic models and matching to large scale structure. The two distributions are very similar in structure given that the Scoccimarro et al. (2001) model has a different normalization.†　Nonetheless, these more sophisticated models for the division of halos into galaxies would greatly improve our more qualitative exploration of the problem.

† When using the Scoccimarro et al. (2001) models to populate the halos from an N-body simulation of the cosmological model in §2 with galaxies, we found that they underestimated the galaxy density by factor of $\simeq 3$.

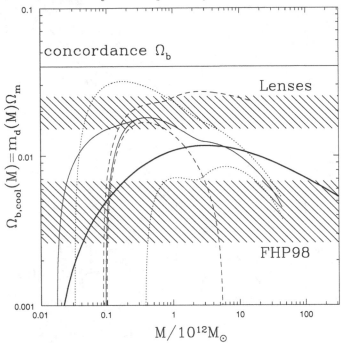

FIGURE 15. Cold baryon densities ($\Omega_{b,\text{cool}} = m_d \Omega_m$). For a range of models of the galaxy formation probability, we use our adiabatic compression models with $\lambda = \bar{\lambda} = 0.05$ and $j_d/m_d = 1$ to convert $v_c(M)$ into an estimate for the cold baryon density. The light solid curves are for a model with our standard high mass cutoff ($\log M_h/M_\odot = 12.5$, $\delta_h = 1$) with (upper curve at low mass) or without (lower curve at low mass) the standard low mass cutoff ($\log M_l/M_\odot = 11.5$, $\delta_l = 3/4$) in the formation probability. The dashed lines show consequences of raising the upper cutoff to $\log M_h/M_\odot = 13$ (bottom) or lowering it to $\log M_h/M_\odot = 12$ (top). The dotted lines show the consequences of incompleteness ($p_0 = 0.5$, upper) and using too low a mass function normalization ($p_0 = 2$, lower) including only the standard high mass cutoff and no low mass cutoff. The heavy solid curve shows the cold baryon fraction estimate by Gonzalez et al. (2000) for the Somerville & Primack (1999) semi-analytic models. The cold baryon density must be less than the total baryon density, $\Omega_{b,\text{cool}} < \Omega_b$, and the total baryon density for the concordance cosmological model is marked with the upper horizontal line. The hatched region labeled FHP98 shows the cold baryon density estimated by Fukugita et al. (1998) scaled to the concordance value of H_0.

We finally return to the problem of the dynamical baryon discrepancy. In Fig. 15 we convert some of the allowed mass-velocity relations into estimates of the cold baryon fraction based on the adiabatic compression model scalings (eqns. 2.1 and 2.2). We also show the equivalent relation Gonzalez et al. (2000) derived from the Somerville & Primack (1999) semi-analytic models. We immediately see that the discrepancy between the estimated baryon content of galaxies (from Fukugita et al. 1998) or the Galaxy (based on the models of Dehnen & Binney 1998) and the cold baryon fraction needed to compress the halos is present whether we use our simple model of §2 and the separation distribution of lenses, full semi-analytic models, or totally non-parametric models.

7. The Future

In this review we examined the relationship between the halo mass function and dynamical probes of it using either the kinematics of galaxies or the separation distribution of gravitational lenses. Although many of the results we discuss are implicit in full semi-analytic models, the approach of examining only the dynamical properties has the advantage of eliminating the dependence of the comparison on luminosity. Dynamical comparisons emphasize the critical role of the cooling baryons in transforming the dynamical structure of halos. The baryonic compression produces a feature in both the distribution of gravitational lens image separations and the local velocity function which is a direct probe of the cold baryon fraction compressing galactic halos. In all our comparisons we find a factor of 2–3 discrepancy between the mass fraction in cold baryons required to explain the data and typical estimates for the cold baryons in galaxies. In a universe with $\Omega_b \simeq 0.04$ we need $0.01 \lesssim \Omega_{b,\text{cool}} \lesssim 0.02$ to compress the halos while typical accountings in galaxies find only $0.005 \lesssim \Omega_{b,\text{cool}} \lesssim 0.01$ in known baryonic components. The difference could be explained by MACHOS (see Alcock, Richer & Sahu in these proceedings), cold molecular gas or even warm gas in some circumstances. It is a part of our general problem that we lose track of most of the baryons at redshift zero (see Tripp in these proceedings).

There are three possible solutions to the dynamical baryon discrepancy. First, it could be imaginary—if you select your preferred ranges appropriately you can essentially eliminate the discrepancy. Second, it could be a problem in our models. The adiabatic compression models are crude approximations for the transformation of the dark matter halos by the baryons, and our models have not properly accounted for the halo multiplicity function. There is also a major debate at present about the consistency of observed rotation curves with the predictions of these standard models (see Burkert, Sancisi & Sanders in these proceedings). It is difficult to adequately address this possibility, since it is currently impossible to compute the final structure of a galaxy starting from the initial halo properties without approximations. Third, the accounting for the baryons in galactic halos may be incorrect. The Fukugita et al. (1998) accounting for the baryons in galaxies included only cold gas and normal stellar/stellar remnant populations, neglecting hot gas (10^6 K), warm ionized gas (10^4–10^5 K), and sub-luminous objects (e.g. MACHOS). While hot gas cannot contribute to the adiabatic compression of the halo, the warm components are both difficult to detect and contribute to the compression.

The most ambitious proposal for the future of the galaxy velocity function is to use its evolution to determine the cosmological equation of state (Newman & Davis 2000). The challenge here is to understand, control, and then eliminate all the sources of systematic error which can mimic or bias the evolutionary effect. Unfortunately, the evolution is subtle and achieving this goal will be difficult. To put these difficulties in some context, many of the sources of systematic uncertainties are identical to those in the (currently unpopular) attempts to determine the cosmological model using gravitational lens statistics. We can see some of the difficulties in our local comparisons between the halo mass function and the velocity function of galaxies. Different routes to deriving the local velocity function (the choice of surveys, luminosity functions, kinematic relations, type distributions $\cdots$), different models for the baryonic mass distribution in galaxies (the distribution of bulge-to-disk ratios, cold baryon fractions, spin parameters $\cdots$) and different models for the connection between halos and galaxies (formation probability, halo multiplicity function $\cdots$) all lead to differences that are comparable to the effects of evolution. Our non-parametric analysis shows that at least two one-dimensional functions of the halo mass are needed, the probability $p_g(M)$ that a halo of mass M forms a galaxy included in the survey, and the average velocity $v_c(M)$ of the resulting galaxy, and both of these functions will themselves be evolving and survey-dependent. However, studying the dynamical evolution of galaxies will yield so much information on the evolution of galaxies, that the experiment will be of enormous value even if systematic problems ultimately prevent it from being used for determining the properties of the background cosmology.

Much of this work was done in collaboration with M. White (§2 and §3) or M. Pahre and E. Falco (§4 and §5), whose contributions and comments were invaluable. CSK is supported by the Smithsonian Institution and NASA grants NAG5-8831 and NAG5-9265.

REFERENCES

BAUGH, C. M., COLE, S., FRENK, C. S. 1996 *MNRAS* **283**, 1361.

BENSON, A. J., ET AL. 2000 *MNRAS* **311**, 793.

BLANCHARD, A., SADAT, R., BARTLETT, J. G., LE DOUR, M. 2000 *A&A* **362**, 809.

BLUMENTHAL, G. R., FABER, S. M., FLORES, R., PRIMACK, J. R. 1986 *ApJ* **301**, 27.

BROWNE, I. W. A., MYERS, S. T. 2000. In *IAU Symposium 201* (eds. A. Lasenby & A. Wilkinson), p. 47. ASP.

BULLOCK, J. S., ET AL. 2000 *Preprint*; astro-ph/9908159.

BULLOCK, J. S., DEKEL, A., KOLATT, T. S., PRIMACK, J. R., & SOMERVILLE, R. S. 2001 *ApJ* **550**, 21.

COHN, J. D., KOCHANEK, C. S., MCLEOD, B. A., & KEETON, C. R. 2001 *ApJ*, **554**, 1216; astro-ph/0008390.

COLE, S., ET AL. 1994 *MNRAS* **271**, 781.

COLE, S. & KAISER, N. 1989 *MNRAS* **237**, 1127.

COLE, S., LACEY, C. G., BAUGH, C. M., & FRENK, C. S. 2000 *MNRAS*, **319**, 167; astro-ph/0007281.

DALCANTON, J. J., SPERGEL, D. N., & SUMMERS, F. J. 1997 *ApJ* **482**, 659.

DE BLOK, W. J. G., MCGAUGH, S. S., BOSMA, A., & RUBIN, V. C. 2001 astro-ph/0103102.

DEIINEN, W. & BINNEY, J. 1998 *MNRAS* **294**, 429.

EFSTATHIOU, G., ELLIS, G., & PETERSON, B. A. 1988 *MNRAS* **232**, 431.

FABER, S. M. & JACKSON, R. E. 1976 *ApJ* **204**, 668.

FALCO, E. E., KOCHANEK, C. S., & MUNOZ, J. A. 1998 *ApJ* **494**, 47.

FLORES, R. & PRIMACK, J. R. 1994 *ApJ* **427**, L1.

FUKUGITA, M., HOGAN, C. J., PEEBLES, P. J. E. 1998 *ApJ* **503**, 518.

FUKUGITA, M. & TURNER, E. L. 1991 *MNRAS* **253**, 99.

GONZALEZ, A. H., ET AL. 2000 *ApJ*, **528**, 145; astro-ph/9908075.

HELBIG, P., MARLOW, D., QUAST, R., WILKINSON, P. N., BROWNE, I. W. A., & KOOPMANS, L. V. E. 1999 *A&AS* **136**, 297.

HERNQUIST, L. 1990 *ApJ* **356**, 359.

JENKINS, A., FRENK, C. S., WHITE, S. D. M., COLBERG, J. M., COLE, S., EVRARD, A. E., & YOSHIDA, N. 2001 *MNRAS* **321**, 372; astro-ph/0005260.

KAUFFMANN, G., COLBERG, J. M., DIAFERIO, A., WHITE, S. D. M. 1999 *MNRAS* **303**, 188.

KAUFFMANN, G., WHITE, S. D. M., GUIDERDONI, B. 1993 *MNRAS* **264**, 201.

KEETON, C. R., FALCO, E. E., IMPEY, C. D., KOCHANEK, C. S., LEHAR, J., MCLEOD, B. A., RIX, H.-W., MUNOZ, J. A., & PENG, C. Y. 2000 *ApJ* **542**, 74; astro-ph/0001500.

KEETON, C. R., MADAU, P. 2000 *ApJ* **549**, L25; astro-ph/0101058.

KEETON, C. R., KOCHANEK, C. S., FALCO, E. E. 1998 *ApJ* **509**, 561.

KEETON, C. 1998 *PhD thesis*, Harvard University.

KITAYAMA T. & SUTO, Y. 1996 *ApJ* **469**, 480.

KOCHANEK, C. S. 1993, *ApJ* **419**, 12.

KOCHANEK, C. S. 1995 *ApJ* **453**, 545.

KOCHANEK,C. S. 1996 *ApJ* **466**, 638.

KOCHANEK, C. S., PAHRE, M. A., FALCO, E. E., HUCHRA, J. P., MADER, J., JARRETT, T. H., CHESTER, T., CUTRI, R., SCHNEIDER, S. E. 2001 *ApJ* **560**, 566; astro-ph/0011456.

KOCHANEK, C. S. & WHITE, M. 2001 *ApJ* **560**, 539; astro-ph/0102334.

KOCHANEK, C. S., PAHRE, M. A., & FALCO, E. E. 2001 *ApJ*, submitted; astro-ph/0011458.

LACEY, C. & COLE, S. 1994 *MNRAS* **271**, 676.

LACEY, C., SILK, J. 1991 *ApJ* **381**, 14.

LI, L-X. & OSTRIKER, J. P. 2000 *preprint*; astro-ph/0010432.

MAHDAVI, A., BOHRINGER, H., GELLER, M. J., & RAMELLA, M. 2000 *ApJ* **534**, 114.

MAOZ, D. & RIX, H.-W. 1993 *ApJ* **416**, 425.

MAOZ, D., RIX, H.-W., GAL-YAM, A., & GOULD, A. 1997 *ApJ* **486**, 75.

MO, H. J., MAO, S., WHITE, S. D. M. 1998 *MNRAS* **295**, 319.

MOORE, B. 1994 *Nature* **370**, 629.

MOORE, B., ET AL. 1998 *ApJ* **499**, L5.

MORTLOCK, D. J. & WEBSTER, R. L. 2000 *MNRAS* **319**, 872; astro-ph/0008081.

NARAYAN, R. & WHITE, S. D. M. 1988 *MNRAS* **231**, 97P.

NAVARRO, J., FRENK, C. S., & WHITE, S. D. M. 1996 *ApJ* **462**, 563.

NEWMAN, J. A. & DAVIS, M. 2000 *ApJ* **543**, L11.

PAHRE, M. A., KOCHANEK, C. S. & FALCO, E. E. 2001, in preparation.

PEACOCK, J. A. & SMITH, R. E. 2000 *MNRAS* **318**, 1144.

PEARCE, F. R., JENKINS, A., FRENK, C. S., COLBERG, J. M., WHITE, S. D. M., THOMAS, P. A., COUCHMAN, H. M. P., PEACOCK, J. A., & EFSTATHIOU, G. 1999 *ApJ* **521**, L99.

PHILLIPS, P. M., BROWNE, I. W. A., WILKINSON, P. N., & JACKSON, N. J. 2000. In *IAU Symposium 201* (eds. A. Lasenby & A. Wilkinson). ASP Conf. Series. ASP; astro-ph/0011032.

PORCIANI, C. & MADAU, P. 2000 *ApJ* **532**, 679.

PRESS, W. & SCHECHTER, P. 1974 *ApJ* **187**, 425.

RIX, H.-W., MAOZ, D., TURNER, E. L., & FUKUGITA, M. 1994 *ApJ* **435**, 49.

SALUCCI, P. 2001 *MNRAS* **320**, L1.

SCOCCIMARRO, R., SHETH, R. K., HUI, L., & JAIN, B. 2001 *ApJ* **546**, 20.

SHETH, R. & TORMEN, G. 1999 *MNRAS* **308**, 119.

SHIMASAKU, K. 1993 *ApJ* **413**, 59.

SIGAD, Y., ET AL. 2000 *preprint*, astro-ph/0005323.

SKRUTSKIE, M. F., ET AL. 1997. In *The Impact of Large Scale Near-IR Sky Surveys* (eds. F. Garzon et al.), p. 187. Kluwer.

SOMERVILLE, R. & PRIMACK, J. 1999 *MNRAS* **310**, 1087.

TULLY, R. B. & FISHER, B. 1977 *A&A* **54**, 661.

TURNER, E. L., OSTRIKER, J. P., & GOTT, J. R. 1984 *ApJ* **284**, 1.

VAN DEN BOSCH, F. C., ROBERTSON, B. E., DALCANTON, J. J., & DE BLOK, W. J. G. 2000 *AJ* **199**, 1579.

VAN DEN BOSCH, F. C. & SWATERS, R. A. 2001 *MNRAS* **325**, 1017.

WAMBSGANSS, J., CEN, R., & OSTRIKER, J. P. 1998 *ApJ* **494**, 29.

WAMBSGANSS, J., CEN, R., OSTRIKER, J. P., & TURNER, E. L. 1995 *Science* **268**, 274.

WHITE, M., HERNQUIST, L., & SPRINGEL, V. 2001 *ApJ* **550**, L129.

WHITE, S. D. M. & FRENK, C. S. 1991 *ApJ* **379**, 52.

WINN, J. N., HEWITT, J. N., & SCHECHTER, P. L. 2001. In *Gravitational Lensing: Recent Progress, Future Goals* (eds. T. Brainerd & C. S. Kochanek), ASP Conf. Series. ASP; astro-ph/9909335.

WU, X.-P., XUE, Y.-J., & FANG, L.-Z. 1999 *ApJ* **524**, 22.

WYITHE, J. S. B., TURNER, E. L., & SPERGEL, D. N. 2001 *ApJ*, **555**, 504; astro-ph/0007354.

Detection of gravitational waves from inflation

By MARC KAMIONKOWSKI[1] AND
ANDREW H. JAFFE[2]

[1]California Institute of Technology, Mail Code 130-33, Pasadena, CA 91125, USA;
kamion@tapir.caltech.edu

[2]Center for Particle Astrophysics, University of California, Berkeley, CA 94720, USA;
jaffe@cfpa.berkeley.edu

Recent measurements of temperature fluctuations in the cosmic microwave background (CMB) indicate that the Universe is flat and that large-scale structure grew via gravitational infall from primordial adiabatic perturbations. Both of these observations seem to indicate that we are on the right track with inflation. But what is the new physics responsible for inflation? This question can be answered with observations of the polarization of the CMB. Inflation predicts robustly the existence of a stochastic background of cosmological gravitational waves with an amplitude proportional to the square of the energy scale of inflation. This gravitational-wave background induces a unique signature in the polarization of the CMB. If inflation took place at an energy scale much smaller than that of grand unification, then the signal will be too small to be detectable. However, if inflation had something to do with grand unification or Planck-scale physics, then the signal is conceivably detectable in the optimistic case by the Planck satellite, or if not, then by a dedicated post-Planck CMB polarization experiment. Realistic developments in detector technology as well as a proper scan strategy could produce such a post-Planck experiment that would improve on Planck's sensitivity to the gravitational-wave background by several orders of magnitude in a decade timescale.

1. What have we learned from the Cosmic Microwave Background?

The past two years have seen spectacular advances in measurements of temperature fluctuations in the cosmic microwave background (CMB; Miller et al. 1999; de Bernardis et al. 2000; Netterfield et al. 2001; Pryke et al. 2001; Hanany et al. 2000; Lee et al. 2001) that have led to major advances in our ability to characterize the largest-scale structure of the Universe, the origin of density perturbations, and the early Universe. The primary aim of these experiments has been to determine the power spectrum, C_ℓ, of the CMB as a function of multipole moment ℓ. Given a map of the temperature $T(\hat{n})$ in each direction $\hat{n}$ on the sky, the power spectrum can be obtained by expanding in spherical harmonics,

$$a_{\ell m} = \int d\hat{n}\, Y_{\ell m}(\hat{n})\, T(\hat{n}), \tag{1.1}$$

and then squaring and summing the coefficients,

$$C_\ell = \frac{1}{2\ell + 1} \sum_m |a_{\ell m}|^2. \tag{1.2}$$

If the map covers a patch of the sky that is small enough to be approximated as a flat surface, the power spectrum can be written in terms of Fourier coefficients:

$$T_{\vec{\ell}} = \int d\hat{n}\, e^{-i\vec{\ell}\cdot\vec{\theta}}\, T(\hat{n}), \tag{1.3}$$

and then

$$C_\ell \simeq \left\langle |T_{\vec{\ell}}|^2 \right\rangle_{|\ell|=\ell}, \tag{1.4}$$

162

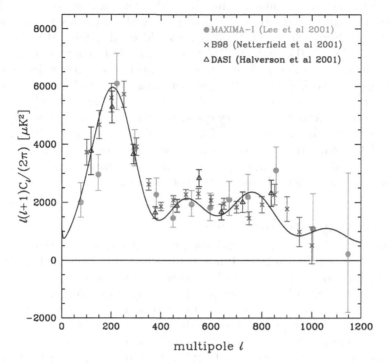

FIGURE 1. Data points on the CMB temperature power spectrum obtained by BOOMERanG, DASI, and MAXIMA. The data indicate unequivocally a peak at $\ell \sim 200$. Moreover, the second and third peaks can be seen in the BOOMERANG and DASI points, and are consistent with the MAXIMA points.

where the average is taken over all Fourier coefficients $\vec{\ell}$ that have amplitude ℓ. Thus, each multipole moment C_ℓ measures, roughly speaking, the rms temperature fluctuation between two points separated by an angle $\theta \simeq (\ell/200)^{-1}$ degrees on the sky.

Recent experiments have sought to determine the power spectrum in the range $50 \lesssim \ell \lesssim 1000$, as structure-formation theories predict a series of bumps in this regime that arise as consequences of oscillations in the baryon-photon fluid in the era before recombination (as indicated by the curves in Fig. 1). The rich structure in these peaks, which can be characterized, e.g. by the precise heights and widths of the peaks, their locations in ℓ, and the heights of the troughs between the peaks, depends in detail on the values of several classical cosmological parameters, such as the baryon density Ω_b (in units of the critical density), Hubble constant h (in units of 100 km/sec/Mpc), matter density Ω_m, and cosmological constant Ω_Λ; on structure-formation parameters such as the amplitude and spectral index of primordial perturbations (Jungman et al. 1996a,b); and on the character of primordial perturbations (e.g. adiabatic, isocurvature, or topological-defects products). In particular, the location of the first peak depends primarily on the geometry of the Universe (parameterized by the *total* density $\Omega_m athrmtot$), and only secondarily on the other cosmological parameters (Kamionkowski, Spergel & Sugiyama 1994). If the Universe is flat, the first peak is expected to occur at $\ell \sim 200$, while if the Universe has a matter density $\Omega_m \sim 0.3$ (as dynamical measurements indicate) but is open (no cosmological constant), then the first peak should be at $\ell \sim 500$.

Experiments that measure the power spectrum in the regime $50 \lesssim \ell \lesssim 1000$ require high sensitivity to detect the CMB temperature variations of roughly one part in 100,000, and they require subdegree angular resolution. Within the past two years, the first high–signal-to-noise high-angular-resolution maps of the CMB have been published by the BOOMERanG (de Bernardis et al. 2000; Netterfield et al. 2001), MAXIMA (Hanany et al. 2000; Lee et al. 2001), and DASI (Pryke et al. 2001) collaborations. The most recent results are shown in Fig. 1. The data show a peak at $\ell \sim 200$ which provides very strong evidence that the Universe is flat (earlier measurements by the TOCO collaboration (Miller et al. 1999) also indicated a first peak at $\ell \sim 200$, but with lower signal-to-noise). The peak structure is also very consistent with growth of large-scale structure from a nearly scale-invariant spectrum of primordial adiabatic perturbations and very *in*consistent with isocurvature or topological-defect alternatives. The peak structure indicated in Fig. 1 is also beginning to provide valuable information about the values of other cosmological parameters (Lange et al. 2000; Balbi et al. 2000).

2. Inflation and gravitational waves

The flatness of the Universe and adiabatic perturbations suggest that we are on the right track with inflation (Guth 1981; Linde 1982; Albrecht & Steinhardt 1982), a period of accelerated expansion in the very early Universe driven by the vacuum energy associated with some new ultra-high-energy physics. In order to solve the horizon problem for which it was initially proposed, inflation predicts that the Universe is flat. Moreover, shortly after inflation was proposed, it was realized that vacuum fluctuations in the inflaton (the scalar field responsible for inflation) would produce a nearly-scale-invariant spectrum of adiabatic perturbations (Guth & Pi 1982; Hawking 1982; Linde 1982; Starobinsky 1982; Bardeen, Steinhardt & Turner 1983). With the advent of these CMB tests, inflation has now had several opportunities to fail empirically, but it has not. Conservatively, these successes are at least suggestive and warrant further tests of inflation.

Perhaps the most promising avenue toward further tests of inflation is the gravitational-wave background. In addition to predicting a flat Universe with adiabatic perturbations, inflation also predicts that quantum fluctuations in the spacetime metric during inflation would give rise to a stochastic gravitational-wave background with a nearly-scale-invariant spectrum (Abbott & Wise 1984). Quantum fluctuations in the spacetime metric can only be affected by gravitational effects which are quantified completely during inflation by the expansion rate $H_m athrminfl$. This is related through the Friedmann equation to the vacuum-energy density V during inflation, $H_{\text{infl}}^2 = 8\pi V/(3m_{\text{Pl}}^2)$, where m_{Pl} is the Planck mass ($m_{\text{Pl}}^{-2} = G$, Newton's constant, in particle-physics units $\hbar = c = 1$). Thus, the amplitude of the gravitational-wave background is fixed entirely by the vacuum-energy density during inflation, which itself should be proportional to the fourth power of the energy scale E_{infl} of the new physics responsible for inflation. The spectrum of gravitational waves depends on the particular inflationary model, but in most models (and certainly in the simplest inflationary models), it is likely to be very close to scale invariant.

These gravitational waves will produce temperature fluctuations at large angles ($\ell \lesssim 1100$) in the CMB (as shown in Fig. 2). The amplitude of their contribution to the CMB temperature quadrupole can be written (Kamionkowski & Kosowsky 1999)

$$\mathcal{T} \equiv 6\, C_2^{\text{TT,tensor}} = 9.2 \frac{V}{m_{\text{Pl}}^4}, \tag{2.1}$$

(where "tensor" refers to gravitational waves, as they are tensor perturbations to the spacetime metric and "TT" refers to the temperature quadrupole). Since the quadrupole measured by $COBE$, $C_2^{\text{TT}} = (1.0 \pm 0.1) \times 10^{-10}$, is most generally due to some combination

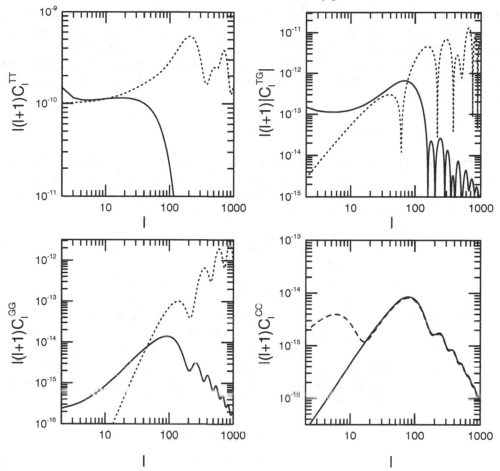

FIGURE 2. Temperature and polarization power spectra from density perturbations (dotted curves) and gravitational waves (solid curves). The absence of a dotted curve for the CC (lower right-hand) panel is due to the fact that density perturbations do not produce a curl component. The solid curves show predictions for a model in which there is no reionization. More realistically some fraction $\tau \sim 0.1$ of the CMB photons will have re-scattered from reionized gas, and this will generate additional polarization power at large angles, as indicated by the dashed curve in the CC panel.

of density perturbations and gravitational waves, we already have an important constraint to the energy scale of inflation: $V^{1/4} \lesssim 2 \times 10^{16}$ GeV (Kamionkowski & Kosowsky 1999; Zibin, Scott, & White 1999).

3. Gravitational waves and polarization

But how can we go further? One might think that improved temperature maps could be used to measure the power spectrum well enough to distinguish the relative contributions of the gravitational-wave and density-perturbation power spectra indicated in the upper left panel of Fig. 2. However, the precision with which the power spectrum can be measured is limited even in an ideal experiment by cosmic variance, the sample variance due to the fact that we have only $2\ell + 1$ independent modes with which to measure each C_ℓ.

Instead, progress can be made with the polarization of the CMB. In addition to producing temperature fluctuations, both gravitational waves and density perturbations will produce linear polarization in the CMB, and the polarization patterns produced by each differ. This can be quantified with a harmonic decomposition of the polarization field. The linear-polarization state of the CMB in a direction $\hat{\mathbf{n}}$ can be described by a symmetric trace-free 2×2 tensor,

$$\mathcal{P}_{ab}(\hat{\mathbf{n}}) = \frac{1}{2} \left(\begin{array}{cc} Q(\hat{\mathbf{n}}) & -U(\hat{\mathbf{n}}) \sin \theta \\ -U(\hat{\mathbf{n}}) \sin \theta & -Q(\hat{\mathbf{n}}) \sin^2 \theta \end{array} \right), \tag{3.1}$$

where the subscripts ab are tensor indices, and $Q(\hat{\mathbf{n}})$ and $U(\hat{\mathbf{n}})$ are the Stokes parameters. Just as the temperature map can be expanded in terms of spherical harmonics, the polarization tensor can be expanded (Kamionkowski, Kosowsky, & Stebbins 1997a,b; Selijak & Zaldarriaga 1997; Zaldarriaga & Seljak 1997),

$$\frac{\mathcal{P}_{ab}(\hat{\mathbf{n}})}{T_0} = \sum_{lm} \left[a^{\mathrm{G}}_{(lm)} Y^{\mathrm{G}}_{(lm)ab}(\hat{\mathbf{n}}) + a^{\mathrm{C}}_{(lm)} Y^{\mathrm{C}}_{(lm)ab}(\hat{\mathbf{n}}) \right], \tag{3.2}$$

in terms of tensor spherical harmonics, $Y^{\mathrm{G}}_{(lm)ab}$ and $Y^{\mathrm{C}}_{(lm)ab}$. It is well known that a vector field can be decomposed into a curl and a curl-free (gradient) part. Similarly, a 2×2 symmetric traceless tensor field can be decomposed into a tensor analogue of a curl and a gradient part; the $Y^{\mathrm{G}}_{(lm)ab}$ and $Y^{\mathrm{C}}_{(lm)ab}$ form a complete orthonormal basis for the "gradient" (i.e. curl-free) and "curl" components of the tensor field, respectively.† Lengthy but digestible expressions for the $Y^{\mathrm{G}}_{(lm)ab}$ and $Y^{\mathrm{C}}_{(lm)ab}$ are given in terms of derivatives of spherical harmonics and also in terms of Legendre functions in Kamionkowski, Kosowsky, & Stebbins (1997b). The mode amplitudes in eq. (3.2) are given by

$$a^{\mathrm{G}}_{(lm)} = \frac{1}{T_0} \int d\hat{\mathbf{n}} \, \mathcal{P}_{ab}(\hat{\mathbf{n}}) \, Y^{\mathrm{G}\,ab\,*}_{(lm)}(\hat{\mathbf{n}}),$$
$$a^{\mathrm{C}}_{(lm)} = \frac{1}{T_0} \int d\hat{\mathbf{n}} \, \mathcal{P}_{ab}(\hat{\mathbf{n}}) \, Y^{\mathrm{C}\,ab\,*}_{(lm)}(\hat{\mathbf{n}}), \tag{3.3}$$

which can be derived from the orthonormality properties of these tensor harmonics (Kamionkowski, Kosowsky, & Stebbins 1997b). Thus, given a polarization map $\mathcal{P}_{ab}(\hat{\mathbf{n}})$, the G and C components can be isolated by first carrying out the transformations in Eq. (3.3) to the $a^{\mathrm{G}}_{(lm)}$ and $a^{\mathrm{C}}_{(lm)}$, and then summing over the first term on the right-hand side of Eq. (3.2) to get the G component and over the second term to get the C component. (In practice, a full likelihood formalism would be used to determine the spectra in the presence of anisotropic, correlated noise, astrophysical foregrounds, and incomplete sky coverage.)

The two-point statistics of the combined temperature/polarization (T/P) map are specified completely by the six power spectra $C^{\mathrm{XX'}}_\ell$ for $\mathrm{X}, \mathrm{X'} = \{\mathrm{T}, \mathrm{G}, \mathrm{C}\}$. Parity invariance demands that $C^{\mathrm{TC}}_\ell = C^{\mathrm{GC}}_\ell = 0$ (unless the physics that gives rise to CMB fluctuations is parity breaking as in Lue, Wang & Kamionkowski 1999; Lepora 1998). Therefore, the statistics of the CMB temperature-polarization map are completely specified by the four sets of moments: C^{TT}_ℓ, C^{TG}_ℓ, C^{GG}_ℓ, and C^{CC}_ℓ.

Both density perturbations and gravitational waves will produce a gradient component in the polarization. However, only gravitational waves will produce a curl component

---

† Our G and C are sometimes referred to as the "scalar" and "pseudo-scalar" components (Stebbins 1996), respectively, or with slightly different normalization as E and B modes (Zaldarriaga & Seljak 1997), although these should not be confused with the radiation's electric- and magnetic-field vectors.

(Kamionkowski, Kosowsky, & Stebbins 1997a; Selijak & Zaldarriaga 1997, but see below). The curl component thus provides a model-independent probe of the gravitational-wave background, and it is thus the CMB polarization component that we focus on here.

4. Detectability of the curl component

If our goal is to detect the polarization signature of gravitational waves, what is the optimum experiment? What is the ideal angular resolution and survey size? What instrumental sensitivity is required? This article will address these questions (although fall a bit short of providing a complete answer).

If we are interested only in the gravitational-wave signature, we can focus on the model-independent curl component of the polarization produced by gravitational waves. In this article, we summarize work reported in Jaffe, Kamionkowski & Wang (2000), a paper that extends the work of Kamionkowski & Kosolowsky (1998) and Lesgourges et al. (2000).† We ask, what is the smallest amplitude of a curl component from an inflationary gravitational-wave background that could be distinguished from the null hypothesis of no curl component by an experiment that maps the polarization over some fraction of the sky with a given angular resolution and instrumental noise? If an experiment concentrates on a smaller region of sky, then several things happen that affect the sensitivity: (1) information from modes with $\ell \lesssim 180/\theta$ (where θ^2 is the area on the sky mapped) is lost;‡ (2) the sample variance is increased; (3) the noise per pixel is decreased since more time can be spent integrating on this smaller patch of the sky.

More concretely, suppose we hypothesize that there is a C component of the polarization with a power spectrum that has the ℓ dependence expected from inflation, as shown in Fig. 2, but an unknown amplitude $\mathcal{T}$. We can predict the size of the error that we will obtain from the ensemble average of the curvature of the likelihood function (also known as the Fisher matrix; Jungman et al. 1996a,b). In such a likelihood analysis, the expected error on the gravitational-wave amplitude $\mathcal{T}$ will be $\sigma_{\mathcal{T}}$, where

$$\frac{1}{\sigma_{\mathcal{T}}^2} = \sum_{\ell} \left(\frac{\partial C_\ell^{\mathrm{CC}}}{\partial \mathcal{T}} \right)^2 \frac{1}{(\sigma_\ell^{\mathrm{CC}})^2}. \tag{4.1}$$

Here, the $\sigma_\ell^{\mathrm{CC}}$ are the expected errors at individual ℓ for each C_ℓ^{CC} multipole moments. These are given by (cf. Kamionkowski, Kosowsky, & Stebbins 1997b)

$$\sigma_\ell^{\mathrm{CC}} = \sqrt{\frac{2}{f_{\mathrm{sky}}(2\ell+1)}} \left(C_\ell^{\mathrm{CC}} + f_{\mathrm{sky}} w^{-1} B_\ell^{-2} \right), \tag{4.2}$$

where $w^{-1} = 4\pi s^2/(t_{\mathrm{pix}} N_{\mathrm{pix}} T_0^2)$ is the variance (inverse weight) per unit area on the sky, $f_{mathrmsky}$ is the fraction of the sky observed, and t_{pix} is the time spent observing each of the $N_{mathrmpix}$ pixels. The detector sensitivity is s and the average sky temperature is $T_0 = 2.73\,\mu\mathrm{K}$ (and hence the C_ℓ^{CC} are measured in units that have been scaled by T_0).

† There is also related work in Kinney (1998); Zaldarriaga, Seljak, & Spergel (1997); Copeland, Grivell, & Liddle (1998) in which it is determined how accurately various cosmological and inflationary parameters can be determined in case of a positive detection. Magueijo & Hobson (1997); Hobson & Magueijo (1996) presented arguments regarding partial-sky coverage for temperature maps analogous to those for polarization maps presented here.

‡ This is not strictly true. In principle, as usual in Fourier analysis, less sky coverage merely limits the independent modes one can measure to have a spacing of $\delta l \gtrsim 180/\theta$. In practice, instrumental effects (detector drifts; "1/f" noise) will render the smallest of these bins unobservable.

The inverse weight for a full-sky observation is $w^{-1} = 2.14 \times 10^{-15} t_{\rm yr}^{-1} (s/200\,\mu{\rm K}\,\sqrt{\rm sec})^2$ with $t_{\rm yr}$ the total observing time in years. Finally, B_ℓ is the experimental beam, which for a Gaussian is $B_\ell = e^{-\ell^2 \sigma_\theta^2/2}$. We assume all detectors are polarized.

The error to $C_\ell^{\rm CC}$ has two terms, one proportional to $C_\ell^{\rm CC}$ (the *sample variance*), and another proportional to w^{-1} (the *noise variance*). There are several complications to note when considering these formulae: 1) We never have access to the actual $C_\ell^{\rm CC}$, but only to some estimate of the spectra; 2) the expressions only deal approximately with the effect of partial sky coverage; and 3) the actual likelihood function can be considerably non-Gaussian, so the expressions above do not really refer to "1σ confidence limits."

Here, we are interested in the detectability of the curl component; that is, what is the smallest gravitational-wave amplitude that we could confidently differentiate from zero? Toy problems and experience give us an approximate rule of thumb: the signal is detectable when it can be differentiated from the "null hypothesis" of $C_\ell^{\rm CC} = 0$. Thus, the ℓ component of the gravitational-wave signal is detectable if its amplitude is greater than

$$\sigma_\ell^{\rm CC} = \sqrt{2/(2\ell+1)} f_{\rm sky}^{1/2} w^{-1} e^{\ell^2 \sigma_b^2}. \qquad (4.3)$$

We then estimate the smallest gravitational-wave amplitude $\mathcal{T}$ that can be distinguished from zero (at "1 sigma") by using Eq. (4.1) with the null hypothesis $C_\ell^{\rm CC} = 0$. Putting it all together, the smallest detectable gravitational-wave amplitude (scaled by the largest consistent with *COBE*) is

$$\frac{\sigma_{\mathcal{T}}}{\mathcal{T}} \simeq 1.47 \times 10^{-17} t_{\rm yr} \left(\frac{s}{200\,\mu{\rm K}\,\sqrt{\rm sec}}\right)^2 \left(\frac{\theta}{\rm deg}\right) \Sigma_\theta^{-1/2}, \qquad (4.4)$$

where

$$\Sigma_\theta = \sum_{\ell \geq (180/\theta)} (2\ell+1) \left(C_\ell^{\rm CC}\right)^2 e^{-2\ell^2 \sigma_b^2}. \qquad (4.5)$$

Results of the calculation are shown in Fig. 3. Plotted there is the smallest gravitational-wave amplitude $\mathcal{T}$ detectable at 3σ by an experiment with a detector sensitivity $s = 10\,\mu{\rm K}\,\sqrt{\rm sec}$ that maps a square region of the sky over a year with a given beamwidth. The horizontal line shows the upper limit to the gravitational-wave amplitude from *COBE*. The curves are (from top to bottom) for fwhm beamwidths of 1, 0.5, 0.3, 0.2, 0.1, and 0.05 degrees. The results scale with the square of the detector sensitivity and inversely to the duration of the experiment.

The sensitivity to the gravitational-wave signal is a little better with an 0.5-degree beam than with a 1-degree beam, but even better angular resolution does not improve the sensitivity much. And with a resolution of 0.5 degrees or better, the best survey size for detecting this gravitational-wave signal is about 3 to 5 degrees. If such a fraction of the sky is surveyed, the sensitivity to a gravitational-wave signal (rms) will be about 30 times better than with a full-sky survey with the same detector sensitivity and duration (and thus 30 times better than indicated in Kamionkowski & Kosolowsky 1998; Kamionkowski & Kosowsky 1999). Thus, a balloon experiment with the same detector sensitivity as MAP could in principle detect the same gravitational-wave amplitude in a few weeks that MAP would in a year. (A width of 200 degrees corresponds to full-sky coverage.)

Since the gravitational-wave amplitude is related to the energy scale of inflation, Fig. 3 determines the inflationary energy scale accessible with any given experiment. Some indication of the range of inflationary models that can be probed with past, current, and future experiments is provided in Fig. 4, adapted from Kinney (1998). The parameter r (y axis) increases with increasing gravitational-wave amplitude, or alternatively, with

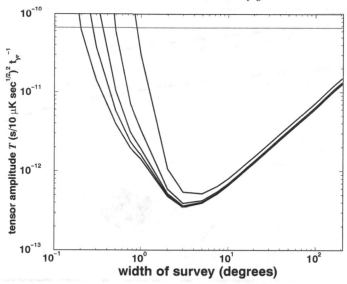

FIGURE 3. The smallest gravitational-wave (tensor) amplitude $\mathcal{T}$ that could be detected at 3σ with an experiment with a detector sensitivity $s = 10\,\mu K\sqrt{\sec}$ that runs for one year and maps a square region of the sky of a given width. The result scales with the square of the detector sensitivity and inversely with the duration of the experiment. The curves are (from top to bottom) for fwhm beamwidths of 1, 0.5, 0.3, 0.2, and 0.1 degrees. The horizontal line shows the upper limit to the gravitational wave amplitude from *COBE*.

the energy scale of inflation. The shaded regions show the where the predictions for various classes of inflationary models (e.g. exponential, power-law, etc.; for more details see Kinney, Dodelson, & Kolb 1998; Kinney 1998). The scored region labeled "COSMIC VARIANCE (no pol)" shows the region of parameter space that would be consistent with a null search for gravitational waves with*out* polarization, while that labeled "PLANCK (with pol)" shows regions of parameter space that would be consistent with a null search for the curl component in the Planck satellite†, an ESA CMB mission scheduled for launch in 2007 (in both cases, it is assumed that $n = 0.95$).

How much could the sensitivity be improved with a post-Planck dedicated polarization experiment? Achieving the Planck sensitivity will be no small feat for experiment, and improvements will require considerable ingenuity. Still, there are prospects for improvements. Very conservatively, a factor-of-3 improvement over Planck's detector sensitivity is plausible. The dark ellipse labeled "$\theta_{fwhm} = 10°$, $\sigma_{pixel} = 1\mu K$" is the error ellipse that could be obtained by a putative experiment with roughly a factor-of-3 improvement to Planck's detector sensitivity, assuming that the true gravitational-wave amplitude and spectral index lie at the center of that ellipse.

There are good reasons to believe that technological developments in the next few years may allow further improvements in detector sensitivity, perhaps of an order of magnitude over that achieved in Planck. As an example, we mention LAMB (Large-Format Array of Microwave Bolometers)‡, a new detector concept that would allow roughly an order-of-magnitude improvement over Planck's detector sensitivity with a much smaller instrument. If we assume that such a detector can be developed and flown in an all-sky survey, the factor-of-ten improvement would allow us to access the regions of parameter space that lie above the line labeled "LAMB" in Fig. 4. If this experiment

† http://astro.estec.esa.nl/SA-general/Projects/Planck/
‡ Bock, J. J., Jones, W., Lange, A. E., & Zmuidzinas, J.; private communication

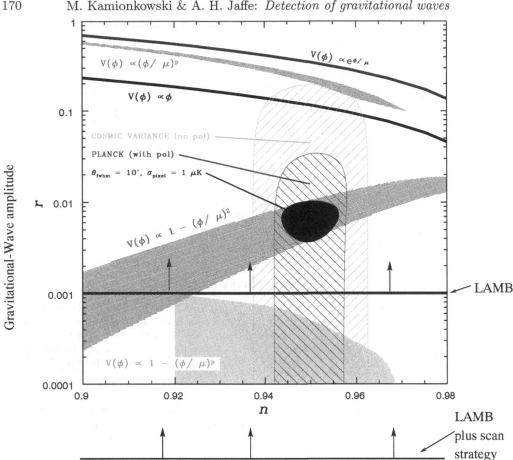

FIGURE 4. Regions in the r-n parameter space occupied by various inflationary models, as well as those regions that could be detected by various CMB experiments. Here r measures the gravitational-wave amplitude (or alternatively, the energy scale of inflation) and n the spectral index for primordial perturbations. Adapted from Kinney (1998). See text for more details.

additionally spent its time surveying a smaller region of the sky, then the regions of inflationary parameter space that could be accessed would be those that lie above the line labeled "LAMB plus scan strategy."

5. Conclusions

We have carried out calculations that will help assess the prospects for detection of the curl component of the polarization with various experiments. Our results can be used to forecast the signal-to-noise for the gravitational-wave signal in an experiment of given sky coverage, angular resolution, and instrumental noise. Of course, the "theoretical" factors considered here must be weighed in tandem with those that involve foreground subtraction and experimental logistics in the design or evaluation of any particular experiment. These usually encourage increasing the signal-to-noise and sky coverage to better isolate experimental systematics.

In contrast to temperature anisotropies which show power on all scales [i.e. $\ell(\ell+1)C_\ell \sim$ const], the polarization power peaks strongly at higher ℓ. Hence the signal-to-noise in a polarization experiment of fixed flight time and instrumental sensitivity may be improved by surveying a smaller region of sky, unlike the case for temperature-anisotropy experiments. The ideal survey for detecting the curl component from gravitational waves is of order 2–5 degrees, and the sensitivity is not improved much for angular resolutions smaller than 0.2 degrees. An experiment with this ideal sky coverage could improve on the sensitivity to gravitational waves of a full-sky experiment by roughly a factor of 30. When coupled with realistic forecasts for improvements in detector sensitivity, we find that an experiment that accesses a very good fraction of the inflationary parameter space (specifically, most of the inflationary parameter space associated with grand unification) is conceivable in the not-too-distant future.

Before closing, we should note that secondary (in the density-perturbation amplitude) effects such as weak gravitational lensing (Zaldarriaga & Seljak 1998) or re-scattering of CMB photons from reionized gas (Hu 2000) may lead to the production of a curl component in the CMB, even in the absence of gravitational waves. However, these secondary effects should be distinguishable from those of gravitational waves, as they produce a curl component primarily at angular scales much smaller than those at which the gravitational-wave signal should show up. Of course, an angular resolution better than that we have suggested here and a survey area a bit larger than we have suggested here may be required to distinguish the gravitational-wave signal from these other sources of a curl component (as well as from foregrounds). A more complete assessment of the impact of these secondary effects on the detectability of gravitational waves is now underway (Kesden & Kamionkowski 2000).

Finally, the CMB polarization will be useful for a wide variety of other purposes in cosmology. For example, detection, and ultimately mapping, of the polarization will help isolate the peculiar velocity at the surface of last scatter (Zaldarriaga & Harari 1995), constrain the ionization history of the Universe (Zaldarriaga 1997), determine the nature of primordial perturbations (Kosowsky 1998; Zaldarriaga & Spergel 1997), probe primordial magnetic fields (Kosowsky & Loeb 1996; Harari, Hayward, & Zaldarriaga 1996; Scannapieco & Ferreira 1997) and cosmological parity violation (Lue, Wang & Kamionkowski 1999; Lepora 1998), and maybe more (see, e.g. Kamionkowski & Kosowsky 1999, for a recent review).

MK was supported in part by NSF AST-0096023, NASA NAG5-8506, and DoE DE-FG03-92-ER40701. AHJ was supported by NSF KDI grant 9872979 and NASA LTSA grant NAG5-6552.

REFERENCES

ABBOTT, L. F. & WISE, M. 1984 *Nucl. Phys. B* **244**, 541.

ALBRECHT, A. & STEINHARDT, P. J. 1982 *Phys. Rev. Lett.* **48**, 1220.

BALBI, M. ET AL. 2000 astro-ph/0005124.

BARDEEN, J. M., STEINHARDT, P. J., & TURNER, M. S. *Phys. Rev. D* **46**, 645.

COPELAND, E. J., GRIVELL, I. J., & LIDDLE, A. R. 1998 *MNRAS* **298**, 1233.

DE BERNARDIS, P. ET AL. 2000 *Nature* **404**, 955.

GUTH, A. H. 1981 *Phys. Rev. D* **28**, 347.

GUTH, A. H. & PI, S.-Y. 1982 *Phys. Rev. Lett.* **49**, 1110.

HANANY, S. ET AL. 2000 *ApJ* **545**, L5.

HARARI, D. D., HAYWARD, J., & ZALDARRIAGA, M. 1996 *Phys. Rev. D* **55**, 1841.

HAWKING, S. W. 1982 *Phys. Lett. B* **115**, 29.

HOBSON, M. P. & MAGUEIJO, J. 1996 *MNRAS* **283**, 1133.

HU, W. 2000 *ApJ* **529**, 12.

JAFFE, A. H. ET AL. 2001 *Phys. Rev. Lett.* **86**, 3475.

JAFFE, A. H., KAMIONKOWSKI, M., & WANG, L. 2000 *Phys. Rev. D* **61**, 083501.

JUNGMAN, G., KAMIONKOWSKI, M., KOSOWSKY, A., & SPERGEL, D. N. 1996 *Phys. Rev. Lett.* **76**, 1007.

JUNGMAN, G., KAMIONKOWSKI, M., KOSOWSKY, A., & SPERGEL, D. N. 1996 *Phys. Rev. D* **54**, 1332.

KAMIONKOWSKI, M. & KOSOWSKY, A. 1998 *Phys. Rev. D* **57**, 685.

KAMIONKOWSKI, M. & KOSOWSKY, A. 1999 *Ann. Rev. Nucl. Part. Sci.* **49**, 77.

KAMIONKOWSKI, M., KOSOWSKY, A., & STEBBINS, A. 1997a *Phys. Rev. Lett.* **78**, 2058.

KAMIONKOWSKI, M., KOSOWSKY, A., & STEBBINS, A. 1997b *Phys. Rev. D* **55**, 7368.

KAMIONKOWSKI, M., SPERGEL, D. N, & SUGIYAMA, N. 1994 *ApJ* **426**, L57.

KESDEN, M. & KAMIONKOWSKI, M. in preparation.

KINNEY, W. H. 1998 *Phys. Rev. D* **58**, 123506.

KINNEY, W., DODELSON, S., & KOLB, E. W. 1998 *Phys. Rev. D* **56**, 3207.

KOSOWSKY, A. 1998 astro-ph/9811163.

KOSOWSKY, A. & LOEB, A. 1996 *ApJ* **469**, 1.

LANGE, A. E. ET AL. 2000 astro-ph/0005004.

LEE, A. T. ET AL. 2001 astro-ph/0104459.

LEPORA, N. 1998 gr-qc/9812077.

LESGOURGUES, J. ET AL. 2000 *A&A* **359**, 414.

LINDE, A. D. 1982a *Phys. Lett. B* **108**, 389.

LINDE, A. D. 1982b *Phys. Lett. B* **116**, 335.

LUE, A., WANG, L., & KAMIONKOWSKI, M. 1999 *Phys. Rev. Lett.* **83**, 1503.

MAGUEIJO, J. & HOBSON, M. P. 1997 *Phys. Rev. D* **56**, 1908.

MILLER, A. D. ET AL. 1999 *ApJ* **524**, L1.

NETTERFIELD, C. B. ET AL. 2001 astro-ph/0104460.

PRYKE, C. ET AL. 2001 astro-ph/0104490.

SCANNAPIECO, E. S. & FERREIRA, P. G. 1997 *Phys. Rev. D* **56**, 4578.

SELJAK, U. & ZALDARRIAGA, M. 1997 *Phys. Rev. Lett.* **78**, 2054.

STAROBINSKY, A. A. 1982 *Phys. Lett. B* **117**, 175.

STEBBINS, A. 1996 astro-ph/9609149.

ZALDARRIAGA, M. 1997 *Phys. Rev. D* **55**, 1822.

ZALDARRIAGA, M. & HARARI, D. D. 1995 *Phys. Rev. D* **52**, 3276.

ZALDARRIAGA, M. & SELJAK, U. 1997 *Phys. Rev. D* **55**, 1830.

ZALDARRIAGA, M. & SELJAK, U. 1998 *Phys. Rev. D* **58**, 023003.

ZALDARRIAGA, M., SELJAK, U., & SPERGEL, D. N. 1997 *ApJ* **488**, 1.

ZALDARRIAGA, M. & SPERGEL, D. N. 1997 *Phys. Rev. Lett.* **79**, 2180.

ZIBIN, J. P., SCOTT, D., & WHITE, M. 1999 *Phys. Rev. D* **60**, 123513.

Cosmological constant problems and their solutions

By ALEXANDER VILENKIN

Institute of Cosmology, Department of Physics and Astronomy, Tufts University, Medford, MA 02155, USA

There are now two cosmological constant problems: (i) why the vacuum energy is so small and (ii) why it comes to dominate at about the epoch of galaxy formation. Anthropic selection appears to be the only approach that can naturally resolve both problems. This approach presents some challenges to particle physics models.

1. The problems

Until recently, there was only one cosmological constant problem and hardly any solutions. Now, within the scope of a few years, we have made progress on both accounts. We now have two cosmological constant problems (CCPs) and a number of proposed solutions. In this talk I am going to review the situation, focusing mainly on the anthropic approach and on its implications for particle physics models. I realize that the anthropic approach has a low approval rating among physicists. But I think its bad reputation is largely undeserved. When properly used, this approach is quantitative and has no mystical overtones that are often attributed to it. Moreover, at present this appears to be the only approach that can solve both CCPs. I will also comment on other approaches to the problems.

The cosmological constant is (up to a factor) the vacuum energy density, ρ_V. Particle physics models suggest that the natural value for this constant is set by the Planck scale M_P,

$$\rho_V \sim M_P^4 \sim (10^{18} \text{ GeV})^4 \ , \tag{1.1}$$

which is some 120 orders of magnitude greater than the observational bound,

$$\rho_V \lesssim (10^{-3} \text{ eV})^4 \ . \tag{1.2}$$

In supersymmetric theories, one can expect a lower value,

$$\rho_V \sim \eta_{\text{SUSY}}^4 \ , \tag{1.3}$$

where η_{SUSY} is the supersymmetry breaking scale. However, with $\eta_{\text{SUSY}} \gtrsim 1$ TeV, this is still 60 orders of magnitude too high. This discrepancy between the expected and observed values is the first cosmological constant problem. I will refer to it as the old CCP.

Until recently, it was almost universally believed that something so small could only be zero, due either to some symmetry or to a dynamical adjustment mechanism. (For a review of the early work on CCP, see Weinberg 1989.) It therefore came as a surprise when recent observations provided evidence that the universe is accelerating, rather than decelerating, suggesting a non-zero cosmological constant (see contributions by Saul Perlmutter and Adam Riess to this volume). The observationally suggested value is

$$\rho_V \sim \rho_{M0} \sim (10^{-3} \text{ eV})^4 \ , \tag{1.4}$$

where ρ_{M0} is the present density of matter. This brings yet another puzzle. The matter density ρ_M and the vacuum energy density ρ_V scale very differently with the expansion of the universe, and there is only one epoch in the history of the universe when $\rho_M \sim \rho_V$. It is difficult to understand why we happen to live in this special epoch. Another, perhaps less anthropocentric statement of the problem is why the epoch when the vacuum energy starts dominating the universe ($z_V \sim 1$) nearly coincides with the epoch of galaxy formation ($z_G \sim 1$–3), when the giant galaxies were assembled and the bulk of star formation has occurred:

$$t_V \sim t_G \ . \tag{1.5}$$

This is the so-called cosmic coincidence problem, or the second CCP.

2. Proposed solutions

2.1. *Quintessence*

Much of the recent work on CCP involves the idea of quintessence (e.g., Zlatev, Wang, & Steinhardt 1999). Quintessence models require a scalar field ϕ with a potential $V(\phi)$ approaching zero at large values of ϕ. A popular example is an inverse power law potential,

$$V(\phi) = M^{4+\beta}\phi^{-\beta} \ , \tag{2.1}$$

with a constant $M \ll M_P$. It is assumed that initially $\phi \ll M_P$. Then it can be shown that the quintessence field ϕ approaches an attractor "tracking" solution

$$\phi(t) \propto t^{2/(2+\beta)} \ , \tag{2.2}$$

in which its energy density grows relative to that of matter,

$$\rho_\phi/\rho_M \sim \phi^2/M_P^2 \ . \tag{2.3}$$

When ϕ becomes comparable to M_P, its energy dominates the universe. At this point the nature of the solution changes: the evolution of ϕ slows down and the universe enters an epoch of accelerated expansion. The mass parameter M can be adjusted so that this happens at the present epoch.

A nice feature of the quintessence models is that their evolution is not sensitive to the choice of the initial conditions. However, I do not think that these models solve either of the two CCPs. The potential $V(\phi)$ is assumed to vanish in the asymptotic range $\phi \to \infty$. This assumes that the old CCP has been solved by some unspecified mechanism. The coincidence problem also remains unresolved, because the time of quintessence domination depends on the choice of the parameter M, and there seems to be no reason why this time should coincide with the epoch of galaxy formation.

2.2. *k-essence*

A related class of models involves k-essence, a scalar field with a non-trivial kinetic term in the Lagrangian (Armendáriz-Picon, Steinhardt, & Mukhanov 2000),

$$L = \phi^{-2}K[(\nabla\phi)^2] \ . \tag{2.4}$$

For a class of functions $K(X)$, the energy density of k-essence stays at a constant fraction of the radiation energy density during the radiation era,

$$\rho_\phi/\rho_{\rm rad} \approx {\rm const} \ , \tag{2.5}$$

and starts acting as an effective cosmological constant with the onset of matter domination. The function $K(X)$ can be designed so that the constant in Eq. (2.5) is $\lesssim 10^{-2}$, thus avoiding conflict with nucleosynthesis, and that k-essence comes to dominate at $z \sim 1$.

This is an improvement over quintessence, since the accelerated expansion in this kind of models always begins during the matter era. Galaxy formation can also occur only in the matter era, but still there seems to be no reason why the two epochs should coincide. The epoch of k-essence domination z_V is determined by the form of the function $K(X)$, and the epoch of galaxy formation z_G is determined by the amplitude of primordial density fluctuations,

$$Q = \delta\rho/\rho \sim 10^{-5} \ . \tag{2.6}$$

It is not clear why these seemingly unrelated quantities should give $z_V \sim z_G$ within one order of magnitude. And of course the old CCP also remains unresolved.

2.3. *A small cosmological constant from fundamental physics*

One possibility here is that some symmetry of the fundamental physics requires that the cosmological constant should be zero. A small value of ρ_V could then arise due to a small violation of this symmetry. One could hope that ρ_V would be given by an expression like

$$\rho_V \sim M_W^8/M_P^4 \sim (10^{-3} \text{ eV})^4 \ , \tag{2.7}$$

where $M_W \sim 10^3$ GeV is the electroweak scale. There have been attempts in this direction (Kachru, Kumar, & Silverstein 1999; Guendelman & Kaganovich 1999; Arkani-Hamed, Hall, Colda, & Murayama 2000), but no satisfactory implementation of this program has yet been developed. And even if we had one, the time coincidence $t_V \sim t_G$ would still remain a mystery.

Essentially the same remarks apply to the braneworld and holographic approaches to CCPs (some of these ideas are discussed in Michael Dine's summary talk).

2.4. *Anthropic approach*

According to this approach, what we perceive as the cosmological constant is in fact a stochastic variable which varies on a very large scale, greater than the present horizon, and takes different values in different parts of the universe. We shall see that situation of this sort can naturally arise in the context of the inflationary scenario.

The key observation here is that the gravitational clustering that leads to galaxy formation effectively stops at $z \sim z_V$. An anthropic bound on ρ_V can be obtained by requiring that it does not dominate before the redshift $z_{\max}$ when the earliest galaxies are formed. With $z_{\max} \sim 5$ one obtains (Weinberg 1987)

$$\rho_V \lesssim 200\rho_{M0} \ . \tag{2.8}$$

For negative values of ρ_V, a lower bound can be obtained by requiring that the universe does not recollapse before life had a chance to develop (Barrow & Tippler 1986),

$$\rho_V \gtrsim -\rho_{M0} \ . \tag{2.9}$$

The bound (2.8) is a dramatic improvement over (1.1) or (1.3), but it still falls short of the observational bound by a factor of about 50. If all values in the anthropic range (2.8) were equally probable, then $\rho_V \sim \rho_{M0}$ would still be ruled out at a 95% confidence level. However, the values in this range are *not* equally probable. The anthropic bound (2.8) specifies the value of ρ_V which makes galaxy formation barely possible. Most of the galaxies will be not in regions characterized by these marginal values, but rather in regions where ρ_V dominates after the bulk of galaxy formation has occured, that is $z_V \lesssim 1$ (Vilenkin 1995; Efstathiou 1995).

This can be made quantitative by introducing the probability distribution as (Vilenkin 1995)

$$d\mathcal{P}(\rho_V) = \mathcal{P}_*(\rho_V)\nu(\rho_V)d\rho_V \ . \tag{2.10}$$

Here, $\mathcal{P}_*(\rho_V)d\rho_V$ is the prior distribution, which is proportional to the volume of those parts of the universe where ρ_V takes values in the interval $d\rho_V$, and $\nu(\rho_V)$ is the average number of galaxies that form per unit volume with a given value of ρ_V. The calculation of $\nu(\rho_V)$ is a standard astrophysical problem; it can be done, for example, using the Press-Schechter formalism.

The distribution (2.10) gives the probability that a randomly selected galaxy is located in a region where the effective cosmological constant is in the interval $d\rho_V$. If we are typical observers in a typical galaxy, then we should expect to observe a value of ρ_V somewhere near the peak of this distribution.

The prior distribution $\mathcal{P}_*(\rho_V)$ should be determined from the inflationary model of the early universe. Weinberg (1997, 2000) has argued that a flat distribution,

$$\mathcal{P}_*(\rho_V) = \text{const} \ , \tag{2.11}$$

should generally be a good approximation. The reason is that the function $\mathcal{P}_*(\rho_V)$ is expected to vary on some large particle physics scale, while we are only interested in its values in the tiny anthropically allowed range (2.8). Analysis shows that this Weinberg's conjecture is indeed true in a wide class of models, but one finds that it is not as automatic as one might expect (Garriga & Vilenkin 2000, 2001).

Martel, Shapiro and Weinberg (1998; see also Efstathiou 1995, Weinberg 1997) presented a detailed calculation of $d\mathcal{P}(\rho_V)$ assuming a flat prior distribution (2.11). They found that the peak of the resulting probability distribution is close to the observationally suggested values of ρ_V.

The cosmic time coincidence is easy to understand in this approach (Garriga, Livio, & Vilenkin 2000; Bludman 2000). Regions of the universe where $t_V \ll t_G$ do not form any galaxies at all, whereas regions where $t_V \gg t_G$ are suppressed by "phase space," since they correspond to a very tiny range of ρ_V. One finds that the probability distribution for t_G/t_V is peaked at $t_G/t_V \approx 1.5$, and thus most observers will find themselves in galaxies formed at $t_G \sim t_V$.

We thus see that the anthropic approach naturally resolves both CCPs. All one needs is a particle physics model that would allow ρ_V to take different values and an inflationary cosmological model that would give a more or less flat prior distribution $\mathcal{P}_*(\rho_V)$ in the anthropic range (2.8).

3. Models with a variable ρ_V

3.1. *Scalar field with a very flat potential*

One possibility is that what we perceive as a cosmological constant is in fact a potential $V(\phi)$ of some field $\phi(x)$ (Garriga & Vilenkin 2000). The slope of the potential is assumed to be so small that the evolution of ϕ is slow on the cosmological time scale. This is achieved if the slow roll conditions

$$M_P^2 V'' \ll V \lesssim \rho_{M0} \ , \tag{3.1}$$

$$M_P V' \ll V \lesssim \rho_{M0} \ , \tag{3.2}$$

are satisfied up to the present time. These conditions ensure that the field is overdamped by the Hubble expansion, and that the kinetic energy is negligible compared with the potential energy (so that the equation of state is basically that of a cosmological constant term.) The field ϕ is also assumed to have negligible couplings to all fields other than gravity.

Let us now suppose that there was a period of inflation in the early universe, driven by the potential of some other field. The dynamics of the field ϕ during inflation are strongly influenced by quantum fluctuations, causing different regions of the universe to thermalize with different values of ϕ. Spatial variation of ϕ is thus a natural outcome of inflation.

The probability distribution $\mathcal{P}_*(\phi)$ is determined mainly by the interplay of two effects. The first is the "diffusion" in the field space caused by quantum fluctuations. The dispersion of ϕ over a time interval Δt is

$$\Delta\phi \sim H(H\Delta t)^{1/2} \ , \tag{3.3}$$

where H is the inflationary expansion rate. The effect of diffusion is to make all values of ϕ equally probable over the interval $\Delta\phi$. The second effect is the differential expansion. Although $V(\phi)$ represents only a tiny addition to the inflation potential, regions with larger values of $V(\phi)$ expand slightly faster, and thus the probability for higher values of $V(\phi)$ is enhanced. The time it takes the field ϕ to fluctuate across the anthropic range $\Delta\phi_{\rm anth} \sim \rho_{M0}/V'$ is $\Delta t_{\rm anth} \sim (\Delta\phi_{\rm anth})^2/H^3$, and the characteristic time for differential expansion is $\Delta t_{de} \sim HM_P^2/V$.

The effect of differential expansion is negligible if $\Delta t_{\rm anth} \ll \Delta t_{de}$. The corresponding condition on $V(\phi)$ is (Garriga & Vilenkin 2001)

$$V'^2 \gg \rho_{M0}^3/H^3 M_P^2 \ . \tag{3.4}$$

In this case, the probability distribution for ϕ is flat in the anthropic range,

$$\mathcal{P}_*(\phi) - {\rm const} \ . \tag{3.5}$$

The probability distribution for the effective cosmological constant $\rho_V = V(\phi)$ is given by

$$\mathcal{P}_*(\rho_V) = \frac{1}{V'}\mathcal{P}_*(\phi) \ ,$$

and it will also be very flat, since V' is typically almost constant in the anthropic range. As we discussed in Section II, a flat prior distribution for the effective cosmological constant in the anthropic range entails an automatic explanation for the two cosmological constant puzzles.

On the other hand, if the condition (3.4) is not satisfied, then the prior probability for the field values with a higher $V(\phi)$ would be exponentially enhanced with respect to the field values at the lower anthropic end. This would result in a prediction for the effective cosmological constant which would be too high compared with observations.

A simple example is given by a potential of the form

$$V(\phi) = \rho_\Lambda + \frac{1}{2}\mu^2\phi^2 \ , \tag{3.6}$$

where ρ_Λ represents the "true" cosmological constant. We shall assume that $\rho_\Lambda < 0$, so that the two terms in (3.6) partially cancel one another in some parts of the universe. With $|\rho_\Lambda| \sim (1\,{\rm TeV})^4$, the slow roll conditions (3.1), (3.2) give

$$\mu \lesssim 10^{-90}M_P \ . \tag{3.7}$$

Thus, an exceedingly small mass scale must be introduced.

The condition (3.4) yields a lower bound on μ,

$$\mu \gtrsim 10^{-137}M_P \ . \tag{3.8}$$

Here, I have used the upper bound on the expansion rate at late stages of inflation, $H \lesssim 10^{-5}M_P$, which follows from the CMB observations.

We thus see that models with a variable ρ_V can be easily constructed in the framework of inflationary cosmology. The challenge here is to explain the very small mass scale (3.7) in a natural way.

3.2. *Four-form models*

Another class of models, first discussed by Brown & Teitelboim (1988), assumes that the cosmological constant is due to a four-form field,

$$F^{\alpha\beta\gamma\delta} = F\epsilon^{\alpha\beta\gamma\delta} \ , \tag{3.9}$$

The field equation for F is $\partial_\mu F = 0$, so F is a constant, but it can change its value through nucleation of bubbles bounded by domain walls, or branes. The total vacuum energy density is given by

$$\rho_V = \rho_\Lambda + F^2/2 \tag{3.10}$$

and once again it is assumed that $\rho_\Lambda < 0$. The change of the field across the brane is

$$\Delta F = q \ , \tag{3.11}$$

where the "charge" q is a constant fixed by the model. Thus, F takes a discrete set of values, and the resulting spectrum of ρ_V is also discrete. The four-form model has recently attracted much attention (Bousso & Polchinski 2000; Donoghue 2000; Feng, March-Russell, Sethi, & Wilczek 2000; Banks, Dine, & Motl 2001) because four-form fields coupled to branes naturally arise in the context of string theory.

In the range where the bare cosmological constant is almost neutralized, $|F| \approx |2\rho_\Lambda|^{1/2}$, the spectrum of ρ_V is nearly equidistant, with a separation

$$\Delta\rho_V \approx |2\rho_\Lambda|^{1/2}q \ . \tag{3.12}$$

In order for the anthropic explanation to work, $\Delta\rho_V$ should not exceed the present matter density,

$$\Delta\rho_V \lesssim \rho_{m0} \sim (10^{-3} \text{ eV})^4 \ . \tag{3.13}$$

With $\rho_\Lambda \gtrsim (1 \text{ TeV})^4$, it follows that

$$q \lesssim 10^{-90} M_P^2 \ . \tag{3.14}$$

Once again, the challenge is to find a natural explanation for such very small values of q.

In order to solve the cosmological constant problems, we have to require in addition that (i) the probability distribution for ρ_V at the end of inflation is nearly flat, $\mathcal{P}_*(\rho_V) \approx$ const, and (ii) the brane nucleation rate is sufficiently low, so that the present vacuum energy does not drop significantly in less than a Hubble time. Models satisfying all the requirements can be constructed, but the conditions (i), (ii) significantly constrain the model parameters. For a detailed discussion, see Garriga & Vilenkin (2001).

4. Explaining the small parameters

Both scalar field and four-form models discussed above have some seemingly unnatural features. The scalar field models require extremely flat potentials and the four-form models require branes with an exceedingly small charge. The models cannot be regarded as satisfactory until the smallness of these parameters is explained in a natural way. Here I shall briefly review some possibilities that have been suggested in the literature.

4.1. *Scalar field renormalization*

Let us start with the scalar field model. Weinberg (2000) suggested that the flatness of the potential could be due to a large field renormalization. Consider the Lagrangian of the form

$$L = \frac{Z}{2}(\nabla\phi)^2 - V(\phi) \ . \tag{4.1}$$

The potential for the canonically normalized field $\phi' = \sqrt{Z}\phi$ will be very flat if the field renormalization constant is very large, $Z \gg 1$.

More generally, the effective Lagrangian for ϕ will include non-minimal kinetic terms (Donoghue 2000; Garriga & Vilenkin 2001),

$$L = \frac{1}{2}F^2(\phi)(\nabla\phi)^2 - V(\phi) \ . \tag{4.2}$$

Take for example $F = e^{\phi/M}$. Then the potential for the canonical field $\psi = Me^{\phi/M}$ is $V(M\ln(\psi/M))$. This will typically have a very small slope if $V(\phi)$ is a polynomial function. It would be good to have some particle physics motivation either for a large running of the field renormalization, or for an exponential function $F(\phi)$ in the Lagrangian (4.2).

4.2. *A discrete symmetry*

Another approach attributes the flatness of the potential to a spontaneously broken discrete symmetry (Dvali & Vilenkin 2001). The main ingredients of the model are: (1) a four-form field $F_{\mu\nu\sigma\tau}$ which can be obtained from a three-form potential, $F_{\mu\nu\sigma\tau} = \partial_{[\mu}A_{\nu\sigma\tau]}$, (2) a complex field X which develops a vacuum expectation value

$$\langle X \rangle = \eta e^{ia} \ , \tag{4.3}$$

and whose phase a becomes a Goldstone boson, and (3) a scalar field Φ which is used to break a discrete Z_{2N} symmetry.

The action is assumed to be invariant under the following three symmetries: (1) Z_{2N} symmetry under which

$$\Phi \to \Phi e^{i\pi/N}, \qquad a \to -a \quad (\text{or } X \to X^\dagger) \ , \tag{4.4}$$

(2) a symmetry of global phase transformations

$$a \to a + \text{const} \ , \tag{4.5}$$

and (3) the three-form gauge transformation

$$A_{\mu\nu\sigma} \to A_{\mu\nu\sigma} + \partial_{[\mu}B_{\nu\sigma]} \ , \tag{4.6}$$

where $B_{\nu\sigma}$ is a two-form. Below the symmetry breaking scales of X and Φ, the effective Lagrangian for the model can be written as

$$L = \eta^2(\partial_\mu a)^2 - \frac{1}{4}F^2 + (\text{effective interactions}) \ . \tag{4.7}$$

The interactions generally include all possible terms that are compatible with the symmetries. Among such terms is the mixing of the Goldstone a with the three-form potential,

$$g\eta^2 \frac{\langle\Phi\rangle^N}{M_P^N}\epsilon^{\mu\nu\sigma\tau}A_{\nu\sigma\tau}\partial_\mu a \ , \tag{4.8}$$

where $g \lesssim 1$ is a dimensionless coupling and I have assumed that the Planck scale M_P plays the role of the ultraviolet cutoff of the theory.

The effect of the mixing term (4.8) is to give a mass

$$\mu = g\eta \frac{\langle \Phi \rangle^N}{M_P^N} \tag{4.9}$$

to the field a. This mass can be made very small if $\langle \Phi \rangle \ll M_P$ and N is sufficiently large. For example, with $\langle \Phi \rangle \sim 1$ TeV, $\eta \ll M_P$ and $N \geq 6$, we have $\mu \ll 10^{-90} M_P$, as required.

Models of this type can also be used to generate branes with a very small charge. In this case a is assumed to be a pseudo-Goldstone boson, like the axion, and the theory has domain wall solutions with a changing by 2π across the wall. The mixing of a and A couples these walls to the four-form field, and it can be shown that the corresponding charge is

$$q = 2\pi g\eta^2 \frac{\langle \Phi \rangle^N}{M_P^N} \ . \tag{4.10}$$

Once again, the anthropic constraint on q is satisfied for $\langle \Phi \rangle \sim 1$ TeV, $\eta \ll M_P$ and $N \geq 6$.

The central feature of this approach is the Z_{2N} symmetry (4.4). What makes this symmetry unusual is that the phase transformation of Φ is accompanied by a charge conjugation of X. It can be shown, however, that such a symmetry can be naturally embedded into a left-right symmetric extension of the standard model (Dvali & Vilenkin 2001).

4.3. *String theory inspired ideas*

Feng et al. (2000) have argued that branes with extremely small charge and tension can naturally arise due to non-perturbative effects in string theory. A potential problem with this approach is that the small brane tension and charge appear to be unprotected against quantum corrections below the supersymmetry breaking scale (Dvali & Vilenkin 2001). The cosmology of this model is also problematic, since it is hard to stabilize the present vacuum against copious brane nucleation (Garriga & Vilenkin 2001).

A completely different approach was taken by Bousso and Polchinski (2000). They assume that several four-form fields F_i are present so that the vacuum energy is given by

$$\rho_V = \rho_\Lambda + \frac{1}{2} \sum_i F_i^2 \ . \tag{4.11}$$

The corresponding charges q_i are not assumed to be very small, but Bousso and Polchinski have shown that with multiple four-forms the spectrum of the allowed values of ρ_V can be sufficiently dense to satisfy the anthropic condition (3.13) in the range of interest. However, the situation here is quite different from that in the single-field models. The vacua with nearby values of ρ_V have very different values of F_i, and there is no reason to expect the prior probabilities for these vacua to be similar. Moreover, the low energy physics in different vacua is likely to be different, so the process of galaxy formation and the types of life that can evolve will also differ. It appears therefore that the anthropic approach to solving the cosmological constant problems cannot be applied to this case (Banks, Dine & Motl 2001).

5. Concluding remarks

In conclusion, it appears that the only approach that can solve both cosmological constant problems is the one that attributes them to anthropic selection effects. In this approach what we perceive as the cosmological constant is in fact a stochastic variable which varies from one part of the universe to another. A typical observer then finds himself in a region with a small cosmological constant which comes to dominate at about the epoch of galaxy formation. Cosmological models of this sort can easily be constructed in the framework of inflation. What one needs is either a scalar field with a very flat potential, or a four-form field coupled to branes with a very small charge. Some interesting suggestions have been made on how such features can arise; the challenge here is to implement these suggestions in well motivated particle physics models. (One attempt in this direction has been made in Dvali & Vilenkin 2001.)

There are also problems to be addressed on the astrophysical side of the anthropic approach. All anthropic calculations of the probability distribution (2.10) for ρ_V assumed that observers are in giant galaxies like ours and identified $\nu(\rho_V)$ in Eq. (2.10) with the density of such galaxies. This, however, needs some justification.† In the hierarchical structure formation scenario, dwarf galaxies could form as early as $z = 10$, and if they are included among the possible sites for observers, then the expected epoch of vacuum domination would be $z_V \sim 10$.

One problem with dwarf galaxies is that if the mass of a galaxy is too small, then it cannot retain the heavy elements dispersed in supernova explosions. Numerical simulations suggest that the fraction of heavy elements retained is $\sim 30\%$ for a $10^9 \, M_\odot$ galaxy and is negligible for much smaller galaxies (Mac Low & Ferrara 1998). The heavy elements are necessary for the formation of planets and of observers, and thus one has to require that the structure formation hierarchy should evolve up to mass scales $\sim 10^9 \, M_\odot$ or higher prior to vacuum domination.

Another point to note is that smaller galaxies formed at earlier times have a higher density of matter. If this translates into a higher density of stars (or dark matter clumps), then we may have additional constraints by requiring that the timescales for disruption of planetary orbits by stellar encounters should not be too short. However, the cross-section for planetary orbit disruption is rather small (comparable to the size of the Solar system), and since close stellar encounters are quite rare in our galaxy, one does not expect a large effect from a modest density enhancement in dwarf galaxies.

An interesting possibility is that disruption of orbits of comets, rather than planets, could be the controlling factor of anthropic selection (Spergel 2001). Comets move around the Sun, forming the Oort cloud of radius ~ 0.1 pc (much greater than the Solar system!). Whenever a star or a molecular cloud passes by, the orbits of some comets are disrupted and some of them enter the interior of the Solar system. Occasionally they hit planets, causing mass extinctions. The time it took to evolve intelligent beings after the last major hit is comparable to the typical time interval between hits on Earth ($\sim 10^8$ yrs), so one could argue that a substantial increase in the rate of hits might interfere with the evolution of observers. There are, of course, quite a few blanks to be filled in this scenario, and at present we are far from being able to reliably quantify the scale of bound systems to be used in the definition of $\nu(\rho_V)$. However, if the anthropic approach is on the right track, then one can *predict* that future research will show the relevant scale to be that of giant galaxies. Finally, I would like to mention the possibility of a 'compromise' solution to CCPs. It is conceivable that the cosmological constant will eventually be determined from the fundamental theory. For example, it could be given by the relation (2.7). This would solve the old CCP. The time coincidence problem could then be solved anthropically if the amplitude of density fluctuations Q is a stochastic variable. With

† I am grateful to David Spergel for emphasizing this to me.

some mild assumptions about the probability distribution $\mathcal{P}_*(Q)$, one finds that most galaxies will form at about the time of vacuum domination (Garriga & Vilenkin 2001).

I am grateful to Mario Livio and Paul Frampton for inviting me to the very interesting meetings where this work was presented and to Gia Dvali, David Spergel and Steve Barr for useful discussions. This work was supported in part by the National Science Foundation and by the Templeton Foundation.

REFERENCES

ARKANI-HAMED, N., HALL, L. J., COLDA, C. & MURAYAMA, H. 2000 *Phys. Rev. Lett.* **85**, 4434.

ARMENDÁRIZ-PICON, C., STEINHARDT, P. J., & MUKHANOV, V. F. 2000 *Phys. Rev. Lett.* **85**, 4438.

BANKS, T., DINE, M., & MOTL, L. 2001 *JHEP* 0101:031.

BARROW, J. D. & TIPLER, F. J. 1986 *The Anthropic Cosmological Principle*. Clarendon.

BLUDMAN, S. 2000 *Nucl. Phys.* **A663-664**, 865.

BOUSSO, R. & POLCHINSKI, J. 2000 *JHEP* 0006:006.

BROWN, J. D. & TEITELBOIM, C. 1988 *Nucl. Phys.* **279**, 787.

DONOGHUE, J. 2000 *JHEP* 0008:022.

DVALI, G. & VILENKIN, A. 2001 Field theory models for variable cosmological constant; hep-th/0102142.

EFSTATHIOU, G. 1995 *MNRAS* **274**, L73.

FENG, J. L., MARCH-RUSSELL, J., SETHI, S., & WILCZEK, F. 2000 Saltatory relaxation of the cosmological constant; hep-th/0005276.

GARRIGA, J., LIVIO, M., & VILENKIN, A. 2000 *Phys. Rev.* **D61**, 023503.

GARRIGA, J. & VILENKIN, A. 2000 *Phys. Rev.* **D61**, 083502.

GARRIGA, J. & VILENKIN, A. 2001 Solutions to the cosmological constant problems; hep-th/0011262.

GUENDELMAN, E. & KAGANOVICH, A. 1999 *Phys. Rev.* **D60**, 165004.

KACHRU, S., KUMAR, J., & SILVERSTEIN, E. 1999 *Phys. Rev.* **D59**, 106004.

MAC LOW, M. & FERRARA, A. 1998 Starburst-driven mass loss from dwarf galaxies: efficiency and metal ejection; astro-ph/9801237.

MARTEL, H., SHAPIRO, P. R., & WEINBERG, S. 1998 *ApJ* **492**, 29.

SPERGEL, D. N. 2001 private communication.

VILENKIN, A. 1995 *Phys. Rev. Lett.* **74**, 846.

WEINBERG, S. 1987 *Phys. Rev. Lett.* **59**, 2607.

WEINBERG, S. 1989 *Rev. Mod. Phys.* **61**, 1.

WEINBERG, S. 1997 In *Critical Dialogues in Cosmology* (ed. N. G. Turok). World Scientific.

WEINBERG, S. 2000 *Phys. Rev.* **D61**, 103505.

ZLATEV, I., WANG, L., & STEINHARDT, P. J. 1999 *Phys. Rev. Lett.* **82**, 896.

Dark matter and dark energy: A physicist's perspective

By MICHAEL DINE

Santa Cruz Institute for Particle Physics, Santa Cruz, CA 95064, USA

For physicists, recent developments in astrophysics and cosmology present exciting challenges. We are conducting "experiments" in energy regimes some of which will be probed by accelerators in the near future, and others which are inevitably the subject of more speculative theoretical investigations. Dark matter is an area where we have hope of making discoveries both with accelerator experiments and dedicated searches. Inflation and dark energy lie in regimes where presently our only hope for a fundamental understanding lies in string theory.

1. Introduction

It is a truism that the development of astronomy, astrophysics, cosmology relies on our understanding of the relevant laws of physics. It is thus no surprise that my astronomy colleagues tend to know more classical mechanics, electricity and magnetism, atomic and nuclear physics than my colleagues in particle theory.

As we consider many of the questions which we now face in cosmology, we must confront the fact that we simply do not know the relevant laws of nature. The public often asks us "What came before the Big Bang?" We usually think of this as requiring understanding of physics at the Planck scale. But at present we can't even come close. Ignorance sets in slightly above nucleosynthesis, and becomes severe by the time we reach the weak scale. Some of the questions which trouble us will be settled by experiment over the next decades; some require new theoretical developments. Needless to say, it is possible that much will remain obscure for a long time.

- GeV scales: QCD is by now a well tested theory, but the phase structure of QCD is not completely understood, and possible first order phase transitions, superconducting phases, strange matter, etc. could be relevant both to astrophysics and cosmology. These questions may be settled by improved lattice gauge calculations, and conceivably by developments at RHIC.

- $T = 100$ GeV-TeV: This is the regime of the weak phase transition. In order to understand this transition, we need experimental information on the Higgs particle, or whatever physics is responsible for the mass of the W and Z. This physics *might* be the origin of the matter-antimatter asymmetry (Cohen et al. 1993). This is physics which will be explored by the Large Hadron Collider at CERN and by a large electron-positron collider, hopefully to be built by an international consortium over the next decade.

- $T = 100$ Gev-TeV: One of the best-motivated candidates for the dark matter is the Lightest Supersymmetric Particle (LSP) expected if nature is supersymmetric. If the supersymmetry hypothesis is correct, we can expect to encounter this particle and its supersymmetric cousins at the Tevatron or LHC and the electron-positron collider. This physics could well be responsible for the matter-antimatter asymmetry.

- $T = 1$ Tev: There have been several suggestions over the past three years that the fundamental scale of physics might lie at the TeV scale (some precursors of these ideas can be found in Rubakov & Shaposhnikov 1983; Dvali & Shifman 1997a,b; Arkani-Hamed et al. 1998; Antoniadis et al. 1998; Randall & Sundrum 1999a,b). In this case, the Planck scale would be so large—and gravity so weak—because there are some very large or highly warped extra dimensions of space. If this hypothesis is correct, there could well be dramatic new phenomena in cosmology just above the temperature of nucleosynthesis.

• $T = 10^{10}$ Gev? The axion is another well-motivated dark matter candidate. It is associated with a symmetry known as a Peccei-Quinn symmetry. 10^{10} GeV might be the scale at which this symmetry is broken. It also might be a scale associated with the dynamics responsible for supersymmetry breaking.

• $T = 10^{15}$ GeV? This could well be the scale of the physics responsible for inflation. Over the past few years, this scale has also emerged as a possible value for the fundamental scale of M theory (Horava & Witten 1996a,b). The two might well be connected (Banks 1999).

• $T > 10^{15}$ GeV? Perhaps physics at these scales holds the explanation of the value of the cosmological constant, and identification of the nature of the dark energy. Perhaps only here lies the physics which resolves the singularity of the big bang, and explains the initial conditions of the universe.

For particle physicists it is extremely exciting to think that there are connections between events in accelerators and our understanding of the history of the universe. Perhaps as important, cosmology can serve as a testing grounds for ideas which are not so readily studied in more conventional experiments. We are probably not going to answer all of the questions which I have listed here soon. But it *is* remarkable that, as we will see, we have hopes of attacking all of them.

2. String or M theory

I will take string theory as a theoretical umbrella in this talk. String theory is a natural framework to talk about all of the issues I have raised above. Indeed, for many of these questions, it is the only framework we have. First, string theory is our only consistent theory of gravity and quantum mechanics. Such a framework is essential if we are to address many of the questions which we face in cosmology. Equally important, string theory encompasses virtually every idea we have for dark matter and energy:

• Low energy supersymmetry, with symmetries like R parity which give rise to a stable, weakly interacting particle.

• Axions: As I will explain further below, string theory is the *only* theoretical context in which we can make sense of the axion hypothesis.

• Cosmological constant: String theory is the only theoretical framework in which we can, even in principle, calculate the cosmological constant. It has realizations all of the various proposed solutions: it has candidates for multiple vacua which might produce an anthropic solution, as mentioned in Vilenkin's talk; it can produce extremely light particles, which realize the other anthropic proposal which Vilenkin mentioned (Vilenkin 2002); and it is "holographic" (to be explained below), so it might offer entirely new solutions.

• Quintessence: String theory is the only context in which we can sensibly discuss the sorts of extremely flat potentials necessary to realize the ideas of quintessence (Peebles & Ratra 1988a,b; Wetterich 1988; Choi 2000).

What is also striking about string theory is that it will allow us to make rather definite statements about many of these ideas. We will see that within our current understanding of string theory, the two anthropic solutions which I mentioned above are implausible.†
Physicists tend to view anthropic explanations of features of physical law with skepticism or worse. Personally, I have for many years thought we might have to contemplate an

† This statement requires some qualification; see (Banks et al. 2001).

anthropic solution of the cosmological constant problem (Weinberg 1987, 1989; Vilenkin 1995). I realized at this meeting that astronomers are more receptive to these ideas than physicists. But what is significant here is that we can potentially use string theory to rule out some anthropic explanations on *scientific* rather than philosophical grounds.

While I will not stress the point here, by similar reasoning, the idea of quintessence is similarly extremely difficult to realize in string theory (Choi 2000).

2.1. *String theory: A quick introduction and survey of recent developments*

What is string theory? At the most simple level, it is just that: a theory of quantized strings. Such a theory is *automatically* a theory which is generally covariant with non-abelian gauge groups. Why? While there has been much progress in understanding these theories, we have at best only a glimpse as to the answer to this question.

More generally, string theory is a framework in which we might hope to address a variety of questions both in particle physics and cosmology. While it is often said to be a theory in ten dimensions, it has solutions with different numbers of dimensions, including four, and

- Standard model gauge interactions
- Repetitive generations (e.g. 3) of quarks and leptons
- Low energy supersymmetry
- Discrete symmetries (R-parity)
- Axions
- Light scalars with very flat potentials (inflatons? quintessence?)
- Exotic possibilities, such as large "compact" dimensions, with dramatic possible implications for particle physics, astrophysics and cosmology.

In the last few years, there have been a number of developments, which are usually grouped together under the heading of duality:

- What were once thought to be several independent string theories have been recognized to be states of one large theory (sometimes called M theory; Witten 1995). Given that the difficulties of quantizing gravity seem so immense, the fact that all previously successful attempts are part of one structure suggests that, just as there is a unique theory of fundamental vector bosons interacting with matter, so there may truly be a unique theory of gravity.
- Many interesting dualities have been understood. For example, many string (and field) theories exhibit an exact electric-magnetic duality (see e.g. J. D. Jackson's book *Classical Electrodynamics*, chapter 6!).
- Many new theoretical tools have been developed, which have permitted the study of quantum aspects of black holes and other real phenomena of quantum gravity. For example, the Beckenstein-Hawking entropy has been understood through the counting of microscopic states (Strominger & Vafa 1996).
- A striking new principle of quantum gravity has been discovered, known as the Holographic principle (for a review, see Bigatti & Susskind 2000): quantum theories of gravity have far fewer degrees of freedom than conventional quantum field theories, such as those of the Standard Model. The number grows like the surface area of the system rather than the volume. This is likely to have profound consequences for the understanding of the question of the cosmological constant (Banks 2000; Cohen et al. 1999; Thomas 2000) and other issues in cosmology (Banks & Fischler 2001).

These developments have provided a number of new insights into longstanding problems. For example, we used to think that the basic scales in string theory would be of order the Planck scale. But with the new developments, we have recognized new possibilities.

• In the strong coupling limit, string theory is best described in terms of an eleven dimensional theory, with gravity propagating in all eleven dimensions, while gauge interactions are confined to ten dimensional walls (Horava & Witten 1996a,b). There is some evidence that this limit is the best suited for describing the real world (Witten 1996). If this idea is correct, the fundamental scale of this theory, the eleven dimensional Planck mass, satisfies:

$$M_{11} \sim 10^{15} \text{ GeV} . \qquad (2.1)$$

The eleventh dimension is curled up, along with the other (more conventional(!?)) six, with radii R_{11} and R given roughly by:

$$R_{11}M_{11} \sim 10 - 30 \qquad R^6 M_{11}^6 \sim 60 . \qquad (2.2)$$

The values of G_N, and the unification of the gauge couplings, give support for this picture.

• Traditionally, in thinking about compactification, one imagined that any extra dimensions were extremely small, of order the Planck mass or unification scale. In recent years, it has been appreciated that extra dimensions might be far larger (Arkani-Hamed et al. 1998; Antoniadis et al. 1998), or could be highly curved (Randall & Sundrum 1999a,b). Either possibility, it has been suggested, might provide an alternative to supersymmetry as a solution to the hierarchy problem. (Prior to these developments, while large extra dimensions had occasionally been suggested, it had not been possible to make sense of them.) These new proposals involve walls or branes in a crucial way, much as in the eleven dimensional limit. The fundamental scale of physics lies at 1 TeV, or so; the smallness of Newton's constant is due to the large size of the extra dimensions, through the relation:

$$G_N = M_{\text{fund}}^8 V_{\text{comp}} \quad M_{\text{fund}} \approx \text{TeV} . \qquad (2.3)$$

The eleven dimensional picture predicts low energy supersymmetry, but also possesses scales of the sort needed to understand the features of inflation (Banks 1999). The large dimension idea predicts:

• Dramatic growth of cross sections for production of Kaluza-Klein modes (e.g. in the process $e^+ + e^- \to \gamma +$ missing energy).

• Cosmology: effects of Kaluza-Klein modes might be important just above nucleosynthesis. For example, some dimensions might be much smaller at early times.

2.2. *What makes string theory hard?*

What makes string theory hard? Why don't we have all the answers? Part of the answer is simply that it is an ambitious theory. It's supposed to explain all the facts of the standard model, *with no parameters*. It is not reasonable to expect all of the answers to fall out so easily. But there are also some specific problems:

• While I said that there are states with desirable properties, there are in some sense too many states. For example, there are states with 11, 10, 9, 8, . . . 4, 3 dimensions; states with or without supersymmetry; states with 1–100s of generations, and so on.

• The classical solutions possess continuous parameters. From the perspective of "low energy" physicists, these are associated with fields. These fields are called moduli. Examples include a field called the dilaton, whose expectation value determines the values of the gauge couplings;, and the radius. We will denote these by $g^{-2}(x^\mu)$ ("dilaton") and $R^2(x^\mu)$, respectively.

• Cosmology: Moduli are candidates for the inflation but they also lead to a set of cosmological difficulties.

• Cosmological constant (more later)

3. Dark matter

Particle physics has provided at least two plausible candidates for dark matter.

• The lightest supersymmetric particle (LSP): requiring that the proton lifetime be long in supersymmetric theories almost inevitably means that the LSP is stable. Supposing that supersymmetry is broken at a scale of order 1 TeV automatically leads to a relic density for this particle of roughly the right order to be the dark matter (the proceedings of this conference feature up to date reviews of both theoretical and experimental issues of dark matter and its detection).

In supersymmetric field theories, it is necessary to postulate discrete symmetries from nowhere to explain the stability of the proton; in string theory, such symmetries are ubiquitous (Green et al. 1986). We heard at this meeting descriptions of ongoing searches for these particles. While the hints from the DARMA experiment are controversial, the 2.6σ discrepancy in $(g-2)_\mu$ provides some cause for optimism that direct evidence for supersymmetry will soon be found (Brown et al. 2001). Indeed, a number of physicists have argued for some time that if the supersymmetry hypothesis is correct, one is likely to see a discrepancy in $g-2$ (Czarnecki & Marciano 2001). Over the next year, further data will be analyzed and the error bars will shrink significantly.

• Axions: The axion is associated with strong CP problem (see, for example, Turner 1990). The axion idea predates the realization that string theory possesses axions by several years (Witten 2000), but it is in string theory that the idea finds a natural home, and indeed it would inevitably would have been discovered there had it not been suggested earlier. In field theory, the Peccei-Quinn solution of the strong CP problem requires that one postulate that nature has a symmetry, which is broken only by tiny quantum effects in the strong interactions. This symmetry must hold so accurately that *extremely* tiny gravitational effects would spoil it, and it has sometimes been argued that this is implausible (Kamionkowski & March-Russell 1992). But in string theory, Peccei-Quinn symmetries of exactly the desired type automatically arise. Prior to the understanding of duality, the Peccei-Quinn scale in string theory was most naturally identified as M_p, so if the axions constituted the dark matter, they were undetectable. With the new understanding, many other possibilities have emerged (Banks & Dine 1997).

4. The problem of the dark energy

As we have heard at this meeting, the evidence for dark energy is mounting. As Professor Livio stressed in his summary, a year ago many astronomers would have doubted the existence of dark energy. Now most, if not totally convinced, are starting to believe it. As Professor Perlmutter remarked, it is particle physicists, especially theorists, who have been his biggest skeptics. As we will see, this is because the result is so surprising. But given that it now seems likely that the data—and its interpretation as dark energy—are correct, it is necessarily a profound clue to the nature of physics at some very different scale.

From the perspective of a particle theorist, the question is: what is the energy density of the vacuum, i.e. of the ground state of whatever is the underlying theory of nature. Obviously it is a tall order to compute this—we need to know the theory—but dimensional analysis suggests we are in trouble. Particle physicists like to describe the cosmological constant, Λ, as a quantity of dimensions of [mass]^4. So

$$\Lambda = M^4 \; , \tag{4.1}$$

where M should be some characteristic mass scale in physics. Is $M = M_p$ (the Planck mass)? $M = M_Z$? $M = m_p$ (the proton mass)? Even in the last case, we would be off by 47 orders of magnitude! Could there be some principle which simply predicts zero? If so, what is the origin of the very tiny observed value?

In string theory, there is some good news with regards to this problem. At the classical level, all of the string vacua I have mentioned have vanishing cosmological constant. While technically easy to describe (Green et al. 1986), this fact is in many ways mysterious. It is not a consequence of symmetries of space time. So this fact represents a striking failure of dimensional analysis, of just the sort we want!

However, even if $\Lambda = 0$ at the classical level of some theory, it is very hard to understand why quantum effects wouldn't generate a huge value for it. The quantum theories of the standard model describe approximately free fields, i.e. they are collections of harmonic oscillators, one for each momentum, spin (and other quantum numbers). The ground state energy of such a theory is then, to lowest order in $\hbar$,

$$\Lambda = \sum (-1)^F \int \frac{d^3k}{(2\pi^3)} \sqrt{\vec{k}^2 + m^2} \quad . \tag{4.2}$$

In general, this expression is very divergent. In our understanding of effective quantum field theories, so successful in describing the standard model, this means that Λ is just a parameter of the theory. This is why physicists (with a few exceptions) traditionally ignored this problem.

On the other hand, at some level, if this is the correct way to think about the vacuum energy, some physics must cut off this integral. What might this be? If nature were exactly supersymmetric, then the bosonic and fermionic contributions to Λ would cancel. If nature is approximately supersymmetric, the integral diverges quadratically, and assuming that physics at the Planck scale provides the cutoff,

$$\Lambda \sim M_{\text{susy}}^2 M_p^2 \approx (10^{10} \text{ GeV})^4 \quad ? \tag{4.3}$$

In string theory, there are no divergences. Since this is a theory of gravity, the calculation of the cosmological constant should be well defined. So this should be a good test of string theory. Does it pass? Is $\Lambda = 0$? 10^{-47}?

Here we have the not so good news: We don't know the answer to this question. All of the string vacua we understand possess moduli at the classical level. These are the light fields with no potential which I referred to above. The expectation values of these fields determine the coupling constant of the string theory, and the masses of various states. Often we can compute a potential for these moduli, but the potential always tends to zero as the coupling tends to zero. Indeed, we have no examples where we can find a stable minimum of the potential in a completely controlled approximation, since, almost by definition, our approximations break down at such a point.

When one discusses moduli, with potentials which fall to zero at infinity, it is natural to consider quintessence. So far, I have spoken of the dark energy as a cosmological constant. As we have heard, many authors have considered the possibility that the dark energy represents some form of quintessence, which I will loosely refer to as the energy of some time-varying field. This is an interesting idea, if only as the equation of state for such a field provides a measure of the quality of future experiments to study the dark energy. It should be noted that, whatever the details of the underlying theory, the mass of the quintessence field today (the second derivative of its potential) can not be significantly smaller than the current horizon size. This is an extremely small number in particle physics units. In other words, not only must the actual value of the present energy density be extremely tiny but so must other quantities.

In string theory, however, it is hard to make sense of the quintessence idea, precisely for the reasons I gave above. The difficulty is that, in examples we can analyze, the scale of the potential is connected to the scale of supersymmetry breaking, which is much too large. So in some sense, one needs to fine tune not only the scale of the potential, but also its derivatives, with extreme precision (Choi 2000). (Since this talk was presented, two papers have appeared noting that there are also serious conceptual issues with quintessence in string theory; Fischler et al. 2001; Hellerman et al. 2001). Quintessence also does not provide a simple explanation of the "why now?" puzzle: the question why the cosmological constant, now, is comparable to the energy density of dark matter. For example, in a simple model (Skordis & Albrecht 2000), this question is resolved by fine-tuning an additional parameter, at the percent or fraction of a percent level. This is not to say that observers should not focus their efforts on measuring w. At the very least, such measurements will give us further confidence that there is a large dark energy component.

Given that quintessence does not fit easily into our current understanding of string theory, let us return to the more conventional cosmological constant idea. If there are stable minima, they are in regions where we can't calculate. Do such stable states exist? Does the cosmological constant vanish, or is it very small for such states (as a consequence of some principle)? Might there be many such states so that we could implement an anthropic solution?

Here there is (reasonably) good news:

• String theory possesses features which allow us to discuss anthropic solutions of the cosmological constant of the type discussed by Vilenkin at this meeting. It has been argued that there might be a very dense set of vacuum states (Bousso & Polchinski 2000; Feng et al. 2001). There can be very flat potentials.

• String theory is *not* like field theory. There is good evidence that it does not possess nearly as many degrees of freedom as field theory. So perhaps the naive quantum estimate we described above is not correct. We might then hope that the classical cosmological constant vanishes, and that the quantum contributions are much smaller than naively expected.

It is interesting that both of theses ideas suggest that the cosmological constant is very small, but not zero.

5. The Anthropic Principle

To the "why now?" question, a number of answers have been offered. Through the years, many authors have noted that an energy scale of order 1 TeV is a natural scale to consider in physics, and that G_N^2 [TeV]^8 is within an order of magnitude of the observed dark energy (in the past, it was argued that it was within such a factor of the limit on the cosmological constant). This is quite impressive, until one remembers that we are indeed trying to explain a coincidence within a factor of two, and that the choice of TeV is very rough. E.g. if it happened that the correct scale was 3 TeV, we would be off by nearly 10^4!

Unfortunately, one can't help but look at the data and conclude that it is pointing us in the direction of some sort of anthropic explanation. Is it possible that if the cosmological constant were much different than observed, the conditions for life, even in its most rudimentary conceivable form, might not be satisfied? I observed at this meeting that astronomers are less afraid to contemplate such a prospect than physicists; they are aware of numerous coincidences in nature which may require such an explanation. In the company of many of my physics colleagues, mentioning the cosmological constant is viewed as barely better than advocating creationism. As Weinberg remarked: "A physicist talking about the anthropic principle runs the same risk as a cleric talking about pornography: no matter how much you say you're against it, some people will think you're a little too interested."

This topic has been reviewed by Vilenkin at this meeting, and I will only add a few remarks. What I have in mind by the anthropic principle in this context is what Weinberg calls the "Weak anthropic principle." The idea is that the universe is vastly larger than what we see within our horizon. In different regions of the universe, the cosmological constant, and possibly other physical constants, take different values. Then, just as people can only live on planets with water, atmospheres, etc. (or, just as fish can only live in water), galaxies/stars/planets/people can only exist in a tiny fraction of the full universe. From galaxy formation, it was originally argued that this hypothesis could not explain a cosmological constant as small as observed (Weinberg 1987; Banks unpublished). More refined arguments give results which may be compatible with what we see (Vilenkin 2002).

As Vilenkin described, there have been a variety of proposals as to how the laws of nature might admit such variation of the parameters. One possibility is that the system has a huge (discrete) number of possible (metastable) ground states, and the distribution of the corresponding energies is nearly continuous (Banks et al. 1991; Bousso & Polchinski 2000; Feng et al. 2001). This quasicontinuous distribution of states has been dubbed a "discretum" (Bousso & Polchinski 2000). Note that the number of states must be enormous. If, for example, the typical scale of the energies is of order 1 TeV, then the number of states must be at least of order 10^{61}. A second possibility is that the universe is permeated by an extremely light field, with the Compton wavelength large compared to the present horizon. As a result this field is currently frozen, but during the inflationary era it fluctuated over a range of values—large enough that in some regions it cancelled any preexisting cosmological constant.

One may not find this mode of explanation appealing, but in some sense it may not matter. Within string theory, one can argue that neither of these proposed explanations is very plausible. Consider, first, the possibility of a very light field, ϕ. The mass of this field has to be smaller than 10^{-50} GeV. There are mechanisms in string theory which could produce a particle this light, but these mechanisms all imply that the maximal value of the field is of order M_p. But in order to cancel off a cosmological constant of order 10^{12} GeV^4, we need $\phi \sim 10^{40} M_p$ or so (Banks & Dine 2001)!

In Banks et al. (1991), it was argued that a peculiar type of axion, known as the "irrational axion," might give rise to a suitable discretum, but subsequent searches have failed to turn up any examples of the required phenomenon in string theory. Four form fluxes in string theory might provide the necessary "discretum" to understand the cosmological constant (Bousso & Polchinski 2000; Feng et al. 2001). Whether this works in detail requires resolving many difficult questions. For example, we need to understand the stability not of just one, but of 10^{120} (or so) states. It also raises the specter that all quantities (the gauge coupling constants, the masses of the elementary particles...) would all be determined anthropically (or alternatively would be random numbers) (Banks et al. 2001). It is hard to imagine that all of the standard model parameters are anthropic, nor do they look like random numbers. So while this idea is the most difficult of the set to rule out, it does not seem particularly promising.

In sum, the remarkable coincidence of the cosmological constant and the present dark matter density is very suggestive of an anthropic explanation. But an anthropic explanation, to be scientific, requires a sensible underlying theory, presumably in the context

of a theory which is capable of making other predictions. So far, we don't have such a
theory.

6. The Holographic Principle

't Hooft and Susskind, from considerations of black hole physics, that in a sensible
theory of gravity in a region of volume V and surface area A, the number of degrees of
freedom must be proportional to A (for a review, see Bigatti & Susskind 2000). The most
familiar piece of evidence for this is the Beckenstein-Hawking entropy formula:

$$S = \frac{G_N A}{4} \ .\tag{6.1}$$

Other features of black hole physics also support this. The fact that in some sense the
information about what is going on in a large volume is encoded in degrees of freedom
residing on the surface is the origin of the term holographic.

There is some evidence that string theory is holographic:

• Naive notions about numbers of degrees of freedom are not correct in string theory. For example, string ground states in smaller dimensions of space time have more
degrees of freedom (suggesting that compactified theories are more "fundamental" than
uncompactified ones).

• String perturbation theory has holographic features: the S matrix seems to be the
crucial observable; C. Thorn argued long ago that the perturbation theory itself has one
the degrees of freedom of a theory in $d-1$ dimensions.

• Two non-perturbative formulations of string theory are known (the "Matrix Model"
and the AdS-CFT correspondence). Both are explicitly holographic.

What might be the implications of this principle for the Cosmological Constant? These
are not clear, but they seem likely to be dramatic, since in the cosmological constant
expression, eqn. 4.2, one might no longer have $V \int d^3k$, but instead a sum over far fewer
degrees of freedom. Are there few enough? The problem is not sufficiently well understood
to say at the present time (Cohen et al. 1999).

This sort of reasoning has lead to even more radical conjectures. In string theory,
there seem to be states with varying numbers of dimensions, and varying amounts of
supersymmetry. Many states with unbroken supersymmetry can be argued to be exact
solutions of the theory. Susskind has suggested that perhaps De Sitter space is not allowed
in string theory. He offers no solid argument, but points to hints based on holography.
Banks proposes that we think very differently about the question of supersymmetry
breaking (Banks 2000). He argues that the number of states in De Sitter space is finite.
Given that recent observations suggest that our universe is De Sitter, what determines
the number of states? Banks proposes that this number is a *parameter*. The cosmological
constant and the amount of supersymmetry breaking are determined by this parameter!
This proposal explains why states with too much supersymmetry might not be viable (and
it is the only proposal which does so). It requires that in holographic theories a different
relation between Λ and the scale of supersymmetry is breaking holds. There is some
reason to believe this might be the case. It requires, however, a very surprising relation
to hold between the cosmological constant and the scale of supersymmetry breaking.

Both ideas are highly speculative, and they are not (yet) supported by a substantial
amount of evidence. But they are suggestive. Indeed, if nothing else, they indicate the
sorts of radical rethinking of many of our basic ideas in physics which may be required
to understand the dark energy.

7. Conclusions

Particle physicists are eager to know the answers to the questions:
- What is the dark matter?
- Is there really dark energy? What is its equation of state?

My experimental colleagues are very interested in dark matter searches, SNAP and ground based proposals to study Type Ia Supernovae. Theorists are hopeful that they have predicted the correct form of the dark matter; they are frantically trying to explain the dark energy. Both are sure to lead to important insights into fundamental law.

I would like to thank many of my colleagues for discussions, and for educating me about the issues raised here, especially Tom Banks, Nathan Seiberg and Leonard Susskind. I would like to thank the organizers and participants in this conference for a very stimulating experience, and particularly to thank Mario Livio. This work was supported in part by the U.S. Department of Energy.

REFERENCES

ANTONIADIS, I., ARKANI-HAMED, N., DIMOPOULOS, S., & DVALI, G. 1998 *Phys. Lett.* **B436**, 257 hep-ph/9804398.

ARKANI-HAMED, N., DIMOPOULOS, S., & DVALI, G. 1998 *Phys. Rev. Lett.* **B429** 263, hep-ph/9803315.

BANKS, T. unpublished.

BANKS, T. 1996; hep-th/9906126.

BANKS, T. 2000; hep-th/0007146.

BANKS, T. & DINE, M. 1997 *Nucl. Phys.* **B505**, 445;hep-th/9608197.

BANKS, T. & DINE, M. 2001; hep-th/0106276.

BANKS, T., DINE, M., & MOTL, L. 2001 *JHEP* **0101**, 031; hep-th/0007206.

BANKS, T., DINE, M., & SEIBERG, N. 1991 *Phys. Lett.* **B273**, 105; hep-th/9109040.

BANKS, T. & FISCHLER, W. 2001; hep-th/0111142.

BIGATTI, D. & SUSSKIND, L. 2000; hep-th/0002044.

BOUSSO, R. & POLCHINSKI, J. 2000 *JHEP* **0006**; hep-th/0004134.

BROWN, H. N., ET AL. 2001 *Phys. Rev. Lett.* **86** 2227; hep-ex/0102017.

CHOI, K. 2000 *Phys. Rev.* **D62**, 043509; hep-ph/9902292.

COHEN, A. G., KAPLAN, D. B., & NELSON, A. E. 1993 *Ann. Rev. Nucl. Part. Sci.* **43**, 27, hep-ph/9302210.

COHEN, A. G., KAPLAN, D. B., & NELSON, A. E. 1999 *Phys. Rev. Lett.* **82**, 4971; hep-th/9803132.

CZARNECKI, A. & MARCIANO, W. J. 2001 *Phys. Rev.* **D64**, 013014; hep-ph/0102122.

DVALI, G. & SHIFMAN, M. 1997 *Nucl. Phys.* **B504**, 127 hep-th/9611213.

DVALI, G. & AND SHIFMAN, M. 1997 PHYS. LETT. **B396**, 64 (E)**B407**, 452, hep-th/9612128.

FENG, J. L., MARCH-RUSSELL, J., SETHI, S., & WILCZEK, F. 2001 *Nucl. Phys.* **B602**, 307; hep-th/0005276.

FISCHLER, W., KASHANI-POOR, A., MCNEES, R., & PABAN, A. 2001; hep-th/0104181.

GREEN, M., SCHWARZ, J., & WITTEN, E. 1986 *Superstring Theory*, Cambridge University Press.

HELLERMAN, S., KALPER, N., & SUSSKIND, L. 2001; hep-th/0104180.

HORAVA, P. & WITTEN, E. 1996a *Nucl. Phys.* **B475**, 94; hep-th/9603142.

HORAVA, P. & WITTEN, E. 1996b *Nucl. Phys.* **B460**, 506; hep-th/9510209.

KAMIONKOWSKI, M. & MARCH-RUSSELL, J. 1992 *Phys. Lett.* **B282**, 137; hep-th/9202003.

PEEBLES, P. J. E. & RATRA, B. 1988l *ApJ* **325**, L17.

PEEBLES, P. J. E. & RATRA, B. 1988b *Phys. Rev.* **D37**, 3406.

RANDALL, L. & SUNDRUM, R. 1999a, hep-ph/9905221.

RANDALL, L. & SUNDRUM, R. 1999b; hep-th/9906064.

RUBAKOV, V. A. & SHAPOSHNIKOV, M. E. 1983 *Phys. Lett.* **125B**, 136.

SKORDIS, C. & ALBRECHT, A. 2000; astro-ph/0012195.

STROMINGER, A. & VAFA, C. 1996 *Phys. Lett.* **B379**, 99; hep-th/9601029.

THOMAS, S. 2000; hep-th/0010145.

TURNER, M. S. 1990 *Phys. Rept.* **197**, 67.

VILENKIN, A. 1995 *Phys. Rev. Lett.* **74**, 846.

VILENKIN, A. 2002; these proceedings.

WEINBERG, S. 1987 *Phys. Rev. Lett.* **59**, 2607.

WEINBERG, S. 1989 *Rev. Mod. Phys.* **61**, 1.

WETTERICH, C. 1988 *Nucl. Phys.* **B302**, 668.

WITTEN, E. 1995 *Nucl. Phys.* **B443**, 85; hep-th/9503124.

WITTEN, E. 1996 *Nucl. Phys.* **B471**, 135; hep-th/9602070.

WITTEN, E. 2000; hep-ph/0002297.